ROTH FAMILY FOUNDATION

Imprint in Music

Michael P. Roth
and Sukey Garcetti
have endowed this
imprint to honor the
memory of their parents,
Julia and Harry Roth,
whose deep love of music
they wish to share
with others.

The publisher and the University of California Press Foundation gratefully acknowledge the generous support of the Roth Family Foundation Imprint in Music, established by a major gift from Sukey and Gil Garcetti and Michael P. Roth.

This publication has been made possible, in part, by funding provided by the AMS 75 PAYS Fund of the American Musicological Society (AMS).

The Composer's Black Box

California Studies in Music, Sound, and Media

James Buhler and Jean Ma, Series Editors

1. *Static in the System: Noise and the Soundscape of American Cinema Culture,* by Meredith C. Ward
2. *Hearing Luxe Pop: Glorification, Glamour, and the Middlebrow in American Popular Music,* by John Howland
3. *Thinking with an Accent: Toward a New Object, Method, and Practice,* edited by Pooja Rangan, Akshya Saxena, Ragini Tharoor Srinivasan, and Pavitra Sundar
4. *Key Constellations: Interpreting Tonality in Film,* by Táhirih Motazedian
5. *Just beyond Listening: Essays of Sonic Encounter,* by Michael C. Heller
6. *Making Stereo Fit: The History of a Disquieting Film Technology,* by Eric Dienstfrey
7. *The Composer's Black Box: Making Music in Cybernetic America,* by Theodore Gordon

The Composer's Black Box

MAKING MUSIC IN
CYBERNETIC AMERICA

Theodore Gordon

UNIVERSITY OF CALIFORNIA PRESS

University of California Press
Oakland, California

© 2026 by Theodore Gordon

All rights reserved.

Library of Congress Cataloging-in-Publication Data

Names: Gordon, Theodore author
Title: The composer's black box : making music in cybernetic America / Theodore Gordon.
Other titles: California studies in music, sound, and media 7.
Description: Oakland, California : University of California Press, [2026] | Series: California studies in music, sound, and media ; 7 | Includes bibliographical references and index.
Identifiers: LCCN 2025019878 (print) | LCCN 2025019879 (ebook) | ISBN 9780520410183 cloth | ISBN 9780520410206 paperback | ISBN 9780520410213 epub
Subjects: LCSH: Electronic music—United States—Philosophy and aesthetics | Electronic composition—History and criticism | Information theory in music | Cybernetics | Electronic musical instruments—Social aspects | Composers—United States—20th century
Classification: LCC ML1092 .G67 2026 (print) | LCC ML1092 (ebook) | DDC 786.7092/2—dc23/eng/20250812
LC record available at https://lccn.loc.gov/2025019878
LC ebook record available at https://lccn.loc.gov/2025019879

GPSR Authorized Representative: Easy Access System Europe, Mustamäe tee 50, 10621 Tallinn, Estonia, gpsr.requests@easproject.com

35 34 33 32 31 30 29 28 27 26
10 9 8 7 6 5 4 3 2 1

Contents

Acknowledgments   vii

    Introduction: What Can a Black Box Do?   1
1. Morton Subotnick and the Composer's Black Box   24
2. Opening Buchla's Box: Manifesting the Musical Mind   49
3. The Patchwork Girl: Pauline Oliveros and the Cybernetic Body   91
4. Alvin Lucier and the Ambiguity of Sound and Signal   124
5. Sun Ra and the Minimoog: Freedom, Discipline, and Opacity   157

    Afterword: What Can a Composer Do?   190

Notes   203
Sources Cited   245
Index   263

Acknowledgments

This book is the product of nearly ten years of reading, listening, thinking, writing, talking, and playing music, and as such there are many people who contributed to the emergence of its final form. The earliest thought and research that would become this book happened at the University of Chicago, where thanks are due to an incredible cohort of friends and scholars, including Nadia Chana, Jessica Peritz, Lindsay Wright, George Adams, Anabel Maler, Ameera Nimjee, William Buckingham, Braxton Shelley, Julianne Grasso, Alican Çamcı, Marcelle Pierson, Daniel Wyche, Cameron Hu, Maria Perevedentseva, and many others. Special thanks are owed to my doctoral advisor Seth Brodsky, along with committee members Jennifer Iverson, Joseph Masco, Steven Rings, and Travis Jackson.

The research for this book took place in libraries and archives across the country, and thanks are due to Jonathan Haim and Jessica Wood at the New York Public Library for the Performing Arts, Mary Ann Quinn at the Rockefeller Archive Center, Carol Lynn Ward-Bamford at the Library of Congress, Janice Braun at Mills College, Heather Smedberg at the UC San Diego Library, Matthew Mehlan at Electronic Sound Studio's Creative Audio Archive, and Natalie Kelsey at Cornell University Library, where Maria Bulla provided essential research assistance. Thanks also to Hannah

Judd for advice on how to navigate the Alton Abraham Collection of Sun Ra at the University of Chicago Library. At Mills College, David Bernstein and Maggi Payne graciously introduced me to the original Buchla Modular Electronic Music System at the Center for Contemporary Music, and answered many queries over the years.

In Berkeley, Perrin Meyer and Barbara Golden generously introduced me to Donald Buchla and many people in his social world. Writing about the 1960s in the 2020s means that many who were there are no longer here, and so I thank Don, along with Bill Maginnis, Tony Gnazzo, Anthony Martin, and Gerd Stern, for their time and conversation before they left this place for another. I also thank Ramon Sender and Morton Subotnick for their time and resolute existence on planet Earth.

Writing this book involved spending lots of time with black boxes—particularly the kinds of boxes that can only really still exist in academic institutions. I thank Margaret Schedel and the late Nicholas R. Nelson for allowing me access to the Electronic and Computer Music Studios at Stony Brook University, where I was able to figure out how, exactly, Pauline Oliveros did or didn't create electronic music with oscillators. At Columbia University, Seth Cluett and Anna Meadors generously facilitated access to the Computer Music Studio, where I was able to spend many hours with their recently restored Buchla Modular Electronic Music System.

The theoretical argument of this book was shaped during my time as a Mellon Postdoctoral Teaching Fellow at Columbia University, where I enjoyed the company and support of another amazing group of friends and colleagues, including Yun Emily Wang, Benjamin Steege, George E. Lewis, Seth Cluett, Ana Maria Ochoa, Ellie Hisama, Eamonn Bell, Marc Hannaford, Dani Dobkin, Jessie Cox, and many more. At Columbia, a seminar based on this book's research was generously supported by the Columbia Center for Science and Society, which reminded me that this project could be of interest outside the relatively cloistered academic disciplines of music studies.

The final version of this book came together during my time at Baruch College, City University of New York, where my colleagues Anne Swartz, Andrew Tomasello, Abby Anderton, and Elizabeth Wollman have been extremely supportive. CUNY has also introduced me to fantastic colleagues and interlocutors including Anne Stone, Scott Burnham, and Eliot Bates at the Graduate Center, Emily Wilbourne and Karen Henson at Queens College, and David Grubbs and Douglas Geers at Brooklyn College.

CUNY's support for this book included a grant from the Faculty Fellowship Publication Program, which supported the research and writing of chapter 5; Michael Gillespie provided extremely constructive feedback. The book has also benefited from several research grants from the Professional Staff Congress, CUNY's union, of which I am a proud member; the final push for completing this book was made possible by a Eugene M. Lang Junior Faculty Research Fellowship from the Weissman School of Arts and Sciences at Baruch College.

The moment when I realized this long-term project was actually a viable book occurred during a manuscript workshop sponsored by the CUNY Gittell Collective (managed by Celina Su, Heath Brown, and Kahina Meziant), for which Deirdre Loughridge, Emily Dolan, Eric Drott, and Fred Turner all graciously agreed to give me five hours of their lives on Zoom and many more reviewing the manuscript. Eric Drott and two anonymous reviewers also reviewed various other versions of the partial and complete manuscript and provided extremely constructive and invaluable feedback.

I could not have asked for a more helpful, constructive, and supportive editorial team at the University of California Press, including my editor Raina Polivka, editorial assistant Sam Warren, and production editor Francisco Reinking. Thanks also to Jon Dertien and Sharon Langworthy for production support and meticulous copyediting.

The biggest thanks are due to the manifold conversations, emails, text exchanges, social media interactions, meals, and sundry activities with the many scholars, musicians, and friends who were willing to entertain my enthusiasm for the period in history, the instruments, the musicians, and the music brought together by this book. In alphabetical order (sorted by a black box, what else?), I thank Adam Leeds, Amy Cimini, Andrei Pohorelsky, Annie Garlid, Benjamin Tausig, Benjamin Piekut, Benjamin Remsen, Brigid Cohen, Catherine Provenzano, Charissa Nobile, Charles Curtis, Charles Eppley, Clara Latham, Daniel Fishkin, Daniel Villegas-Vélez, David Gutkin, David Kant, David Novak, Douglas Barrett, Eve Essex, Ezra Teboul, Keith Fullerton Whitman, Kerry O'Brien, Luciano Chessa, Madison Heying, Marcia Bassett, Mark Mahoney, Matt Mehlan, Michael Gallope, Michelle Yom, the MEMS Project, Patrick Nickleson, Richard

Smith, Robert Gluck, Ryan Dohoney, Sam Kulik, Sam Pluta, Sandy Gordon, Scott Wollschleger, Sumanth Gopinath, Todd Barton, Tom Erbe, Will Robin, and many more.

Susan Barker, Josh Gordon, and Joel Orenstein supported me every step of the way. I dedicate this book to Hannah Burnett, my partner in life and thought, without whom none of it would have been possible, and to Oscar Burnett, who reminds us to be here now.

Introduction

WHAT CAN A BLACK BOX DO?

In early 1970, the Nigerian writer Tam Fiofori began to work on a cover story for *DownBeat* magazine about the emergence of the synthesizer as a new musical instrument for the information age. The previous three years had seen an explosion of media attention to this instrument, even as most people, including musicians, didn't understand what it was, how it worked, or even what, exactly, it did. To most people, including musicians, the synthesizer remained a black box—something that could be observed, but whose internal complexity was functionally unknowable, masked behind an overwhelmingly complex interface of knobs, switches, lights, and plugs. By the time Fiofori began working on his story, the synthesizer itself was not difficult to observe; Wendy Carlos's October 1968 album *Switched-On Bach*, created with a custom Moog synthesizer collaboratively designed by the engineer Robert Moog and Carlos herself, had spent the previous fifty-nine weeks on the *Billboard* pop album chart, spawning copycat albums that ranged in style from blues to Bacharach. But although the synthesizer's sounds were not hard to hear, it was still unclear to most listeners how, exactly, the instrument worked. Despite assurances that the "Moog" was an eponymous musical instrument just like a Stradivarius violin or a saxophone, its bewildering interface remained illegible to most musicians, producing wildly fecund fantasies of what this black box could do. Introducing the Moog synthesizer to a bemused youth audience in 1969, Leonard Bernstein declared that the instrument "could do anything but stand up and take a bow."[1]

Seeking to demystify the synthesizer, Fiofori went directly to Moog to ask: "What special advantage does the instrument put at the disposal of the musician in terms of how he [sic] actually creates new sounds and shapes these sounds?" After describing "traditional instruments" as those that convert energy generated by humans into sound, through plucking, blowing, or bowing, Moog differentiated his novel synthesizer: "In electronic equipment, it's the electronic circuits that make the sound and you can assume the role of shaper. You can devote all your energy to shaping sounds."[2] The term *energy* is used two ways in Moog's response, subtly effecting a consequential transformation from power to information. To Moog, with traditional instruments, energy is transferred kinetically from human fingers or lungs to the instrument's strings or pipes, producing sound; with electronic instruments, energy is transferred electronically from the human mind to the instrument's interface, "shaping" the sound rather than physically producing it.

From this transformation, Moog concluded, the synthesizer promised a fundamental shift in the role of the human musician. "Up until now," Moog observed, "sound has always been sort of architectural. The musician has had to construct the piece of music note by note; he's had to play one note, then another, and so on." Now, with his synthesizer, "you can regard sound as a sort of sculptural activity, as a mass, and put contours and shapes into it." Such "sculpting" would take place not through the musician's hands, but through their mind. Indeed, Moog began his interview with Fiofori by claiming that "right now, with the equipment that is available, you can make just about any sound that you can imagine."[3] To Fiofori's question, "How do you see the new areas of music the synthesizer has opened up?," Chris Swansen, a musician employed by Moog at his factory in Trumansburg, New York, responded: "The possibilities are infinite, in the sense that it is a tool to harness the mind of the musician in realizing in sound [. . .] whatever sound his [sic] mind likes, whatever forms he likes."[4] Among those in Moog's orbit, there seemed to be a consensus that the synthesizer would allow individual human beings to better express themselves through sound—in Swansen's words, creating "an infinite number of new musics, each one representing the personality of the creator."[5]

That contemporary electronics technologies would allow an individual human to finally realize their own individual music, comprised of individually imagined sounds organized in individually shaped compositions, was a vision shared not only by users of Moog synthesizers, but also by users of equipment made by Donald Buchla, a contemporaneous designer of electronic musical instruments in the San Francisco Bay Area. In 1963 or 1964, Buchla had been asked by the composer Morton Subotnick to create what Subotnick imagined as a "music easel" with which a composer could immediately "paint" music, bypassing the complexities of creating electronic music with oscillators and magnetic tape. Unlike Moog's system, which gave its users the option of using a musical keyboard to control the sculpting of sound, Buchla's system was initially built to control specific technological devices in the studio in which it was housed, the San Francisco Tape Music Center (SFTMC), and thus presented users with two specialized "touch plates" in addition to the knobs, switches, lights, and various other controls also present in Moog's instruments. With the "Buchla Box," as it was affectionately known, Subotnick imagined a transformation of the human creation of music from a social endeavor to what he called "music as a studio art," in which an individual human would vertically integrate the activity of music-making as "an instrument maker, performer, listener and conductor."[6] For Subotnick, the absence of a musical keyboard in Buchla's system meant that the musician would be unburdened by the historical legacies of keyboard instruments such as the piano or the organ, furthering the unique individualism of their creations.

Despite the differences between Moog's and Subotnick's electronic instrument design philosophies, however, the goals of their musical ideologies were essentially the same: to liberate the individual human subject by allowing them to express their innermost creative self through electronic sound. Both Moog's and Subotnick's instruments were imagined as tools that could sonify and extend an individual human's imagination through electricity, producing their own personal music. Electronic musical instruments promised a radical transformation of music from a physical to an informational endeavor; although electronics ultimately relied on the physical oscillation of electrons in various wave shapes, frequencies, and amplitudes, these oscillations were understood to be informational rather than merely physical. That such information-carrying electronics could immediately "harness the mind of the musician," as Swansen put it, relied on a novel conception of the mind as similarly informational and electronic. Just as the neck, ribs, and waist of the violin metaphorized the physical contours of the human body, so, apparently, did electronic musical instruments metaphorize the informational contours of the human mind.

But just as some musicians imagined novel electronic musical instruments as emancipating the individual human mind to express the latent music embedded deep within the human subject, others drew fundamentally different conclusions from their encounters with such instruments. Fiofori's *DownBeat* feature also included an essay written by the jazz composer Sun Ra, who had first played a large, modular Moog synthesizer system in 1969 and who would soon become the first musician to own and perform with a prototype keyboard synthesizer called the Minimoog, in July 1970. Rather than allowing for each musician to individually emancipate themselves, Ra saw electronic musical instruments as heralding a different social order of future musicality: "[The synthesizer] is a challenge to the music scene. Perhaps it may galvanize musicians into harmony and harmony units and they will become exponents of harmonic brotherhoods, banded together at last to achieve ensemble presence and worth according to the greater standards of the universe [. . .] Some musicians may fear the synthesizer. It is awesome even in its primary development; but this is a bold, aggressive age, pioneering in a sense beyond known frontiers."[7]

Through the synthesizer, Ra imagined a music transformed into a "language of impression vibrations," themselves "magnified reiterations" of sound itself, a feedback loop of cosmological musical signal revealing itself through electronic circuits. Imagining electronic music as an "intercession," a prayer to a cosmological authority, Ra proposed a different kind of "sound world" built not from individual expression but from "harmonic brotherhoods," ensembles of humans and perhaps instruments, which would accord with the "greater standards of the universe."[8] To Ra, those standards could only be revealed through the "reiteration dimension" of performing with the instruments themselves. While musicians such as Subotnick and Swansen valorized the inputs and outputs of instruments such as the Buchla Box or the Moog synthesizer—the inputs of "creativity" or "imagination," and the outputs of electronic sound—Ra valorized the reiteration of "permutated sound" that operated within the instrument itself.

In both approaches, the electronic musical instrument was conceptualized as an apparatus that took in, processed, and output electronic signal, which could behave like sound or also something more than sound—something like cognition, information, or energy, from the scale of the human to the scale of the cosmos. Such signal was, in both approaches, underdetermined, allowing musicians to observe the same sonic output and attribute it to a multiplicity of different origins. In this sense, a new electronic musical instrument like a Moog or Buchla system behaved like what contemporaneous engineers called a *black box*: a device with observable inputs and outputs that encapsulated and veiled complex, nonlinear processes, rendering its internal workings profoundly opaque.

"THE COMPOSER'S BLACK BOX"

The Composer's Black Box begins with the metaphor of the "black box," borrowing its name from an interview with Subotnick in which he described the instrument he began imagining in the early 1960s specifically as a "composer's black box."[9] Such a black box, to Subotnick, would be a "personal tool for creating music/art with sound, more of an analog computer than a musical instrument."[10] With his composer's black box, Subotnick imagined, "I could create and perform in my studio, and it would come out as a sound piece, which was at once a musical creation and a performance."[11] Simplifying the complex procedures of creating electronic music into a schema of input and output, this black box would take as input a performance of musical creativity, and through its opaque internal workings, output sound. The ambiguity of the terms in this metaphorical schema was convenient for Subotnick, since he had virtually no knowledge of electrical engineering or electronic technology; yet this lack of technical knowledge did not prevent him from developing a powerful fantasy about both the technical and social consequences of his imagined black box. As he recalled later in his life, in 1962 he "had begun to imagine this electronic music easel as a tool for any person who wanted to be creative with sound, to be able to afford it, and to have it in his or her own home. The transistor had arrived, and most of us knew that the consequences of that foretold that electronics were destined to be affordable by all."[12]

Subotnick's vision of a future in which every individual homeowner could own their unique musical black box and use it to create their own personal music, even without musical training, was emblematic of a much larger trend in American social, technical, and political thought in the 1960s. As the historian of science Ronald R. Kline has explored, throughout that decade, "futurists, policymakers, journalists, social scientists, and humanists started writing about the coming of a new era based on computers and communications technology," leading to a nearly immediate and pervasive characterization of that decade and subsequent decades as the "information age."[13] Building on research by the historian David Nye, Kline points out that the 1960s became an origin point for a utopian technological narrative, which "typically emerges at the start of a foundation story and makes extravagant claims that a new technology will bring enormous social and economic benefits."[14] The composer's black box, to Subotnick, was an emblem of such a utopian narrative. In Subotnick's imagined future, musical black boxes would deskill and democratize musical creativity, allowing each individual human to both specify their own unique musical instrument and create their own unique music, all through the novel and opaque flow of electronic signal through their very own black box.

Yet what exactly would be input into the box, how that input would be transformed into an electronic signal, what would happen to that signal inside the box, and how that signal would eventually be output as sound were only imagined in broad generalities. For musicians who encountered such black boxes materially in the form of novel electronic instruments, or who encountered cases of black box thinking more generally, the ambiguity of the black box metaphor produced differing conceptualizations of every entity in the system: human, instrument, and the flowing signal that connected them in a circuit. For Subotnick (and also for Moog and Swansen), the black box was imagined to be an electronic device that would more easily and immediately extend some essential internal element of the individual composer's mind. But if the black box was truly opaque, its internal workings unknown and behavior only understandable through experimental observation, could a human user truly ever be fully in control of its output? What would become of the agency and identity of the human entangled in this black-boxed system?

For Sun Ra, musical black boxes such as his prototype Minimoog did not allow for an individual human being to express themselves more freely or immediately. After playing his Minimoog for nearly two years, Ra described his performances with electronic instruments not as expressing his individuated self, but rather as "demonstrating infinity, because that's the way the Universe is, and it keeps demonstrating that . . . the precision and discipline of it."[15] Ra's position on the question of individual liberation reflected his own social and political experience in mid-century America, which was markedly different from Subotnick's. While Subotnick's racial identity remained unmarked throughout his successful career in American musical, academic, and nonprofit institutions, Ra's marked racial identity meant that his social, economic, and political experience in mid-century America were marked by the discrimination and violence of anti-Black racism. Rather than develop an optimistic view of the future of musical and social life in America, Ra developed a pessimism about Black social life on planet Earth, seeking out alternate planes of existence that would transcend what he experienced as the failure of liberal political ideology in America.

In a 1971 radio interview, Ra summarized his position on freedom:

> Now they've been read up on the word freedom, which is what the white people have been selling. Now when they deal with freedom, they can't play my music, because I don't care anything about freedom. My whole thing is based on discipline and precision, because that's what nature does. There's no freedom in nature. You have your particular place, and you do what you're supposed to do. As far as I'm concerned, the only freedom that I've ever seen that Black people get has been over in the cemetery, and the only peace I've ever seen them get is that 'rest in peace' that the man reserved for them.[16]

These two poles—Subotnick's view of the black box as freeing the individual to better express themselves, and Ra's view of the black box as negating individual freedom and demonstrating the discipline of the universe—characterize opposing poles of the conceptual space in which *The Composer's Black Box* explores musicians' encounters with the scientific concepts and technological objects of the newly developing information age in the 1960s. Within this space, the book explores five case studies of musicians' encounters with contemporary American technoscience and the differing consequences that such encounters had not only for musical aesthetics, but also for social and political conceptions of the musicking human subject. In each case, encounters with contemporary technoscience, specifically with the broad categories of information theory and cybernetics, galvanized musicians to explore questions of musical subjectivity—the relationships between the human body, the human mind, musical instruments, communication, sensation, sound, agency, and freedom.

MAKING MUSIC IN CYBERNETIC AMERICA

The metaphor of the black box was, and remains, one of the most pervasive metaphors in systems engineering. It allows for an engineer to understand the behavior of one part of a system, or even the entire system itself, without needing to understand how it works. As the historian of science Philipp von Hilgers has explored, the black box metaphor has at least two origins, both as "a physical thing within a specific historical constellation, and [...] an epistemic object at the center of an interdisciplinary debate over a new research paradigm" of cybernetics.[17] The term first began to circulate to describe cavity magnetrons, small devices capable of generating microwaves developed in Great Britain to aid in equipping military aircraft with radar capability—a massive strategic advantage that was to be safeguarded at all costs. These cavity magnetrons were doubly opaque; not only were they secretively transported in black metal deed boxes, but at their core was a massive copper ingot with inaccessible internal cavities, after which the device was named. *Black box*, in its military origin, came to refer to a device that purposefully attempted to hide its internal functioning to avoid being obtained and understood by an enemy.

Upon the arrival of these black boxes at the Massachusetts Institute of Technology (MIT) in 1942, however, the term *black box* also contributed to a developing scientific discipline that the mathematician and philosopher Norbert Wiener would later come to call "cybernetics," or the study of self-governing systems. Such systems, either in part or in whole, could be conceptualized as black boxes; the working of their internal mechanisms was immaterial to the study of their behavior and how their inputs and outputs could relate to each other using the concept of feedback. To Wiener, cybernetics sought to theorize "the entire field of control and communication theory, whether in the machine or the animal."[18] The historian of science Peter Galison has traced Wiener's development of cybernetics to his work at MIT on an antiaircraft predictor, a device designed to track an enemy aircraft's position, predict its trajectory, and ultimately aid in the accuracy of shooting it down. As Galison has argued, in developing a device that predicted an enemy pilot's future behavior, Wiener's project quickly grew in scope to encompass not only the behavior of the enemy, but also the Allied operator of the device itself, and thus to the human mind in general, conceptualized as a "vast array of human proprioceptive and electrophysiological feedback systems."[19] As Galison writes, "The AA predictor, along with its associated engineering notions of feedback systems and black boxes, became, for Wiener, the model for a cybernetic understanding of the universe itself."[20]

Although Wiener was only one of a constellation of scientists from different disciplines who began to study connections between behavior, communication, systems theory, and information theory in the 1940s, the publication of his 1948 book *Cybernetics: or, Control and Communication in the Animal and the Machine* gave a name to one of this interdisciplinary milieu's central concepts of a self-governing system that controlled itself through the feedback of informational signal. Such a system, whether mechanical or biological, was understood by Wiener as operating not merely mechanically, but informationally and statistically. In short, as Kline summarizes, "the basic analogy of cybernetics" was that all systems—mechanical, biological, or otherwise—could be described as using "information-feedback paths to adapt to their environment."[21] Such systems were conceptualized as having inputs and outputs that were often nonlinear, meaning that their output did not relate in a directly proportional way to their input. For Wiener, any "as-yet unanalyzed nonlinear system" was defined as a black box; its inputs and outputs were observable, even as its internal workings were opaque. The black box metaphor could apply to anything from a cavity magnetron to the human nervous system itself, which Wiener likened to an "ultra-rapid computing machine." Understood by Wiener as an opaque, nonlinear system with observable inputs and outputs, "the synapse is nothing but a mechanism for determining whether a certain combination of outputs from other selected elements will or will not act as an adequate stimulus for the discharge of the next element, and must have its precise analogue in the computing machine."[22]

Although the science and technology of cybernetics may seem far removed from the ostensibly cultural domain of music, throughout the 1950s and 1960s, its central metaphors quickly came to saturate many aspects of American life well outside the sciences. Systematicity and self-governance through feedback could be seen everywhere and anywhere by both scientists and nonexperts; the universalism of the discourse allowed it to be embraced even as its scientific specificity and rigor was diluted. As Geoffrey Bowker has argued, the language of cybernetics "operated as a kind of legitimacy exchange"; cybernetic claims from one discipline could point to cybernetic claims from another discipline that referenced the first, allowing cybernetics to quickly jump between academic disciplines. "Even if the language were presumed to have no content whatever," Bowker observed, "it could still provide an opportunity to use interesting words and so make useful associations."[23] Although Bowker was referring specifically to scientific disciplines, Christina Dunbar-Hester has pointed out that "cybernetics was more than a universal language within the sciences," especially given the language's rapid adoption among experimental and electronic musicians.[24] Even by the early 1960s, terms such as *information*, *feedback*, *noise*, and *black boxes* had become so commonplace that even a composer such as Subotnick, who had no formal scientific education, could easily use them to describe an imagined "analog computer for music" that would function as a "composer's black box."[25]

Subotnick was far from the only musician to draw ideas from the expansive interdiscipline of cybernetics in the 1960s. Indeed, beginning with the emergence of cybernetics in the 1940s, composers, instrument designers, and technicians across the globe began to imagine new compositional techniques and musical instruments that utilized concepts from the many disciplines that cybernetics embraced in its broad purview, including information theory, statistical analysis, cognitive science, and behavioral psychology. In Europe, theorists such as Werner Meyer-Eppler and Abraham Moles, both of whom were trained as physicists and had extensive experience in electrical and acoustic engineering, developed robust theories that attempted to describe human auditory perception, human vocal production, and musical aesthetics through information theory. As Jennifer Iverson has explored at length, Meyer-Eppler's work on experimental phonetics informed his development of techniques and technologies for the production of *Elektronische Musik* at the West-Deutscher Rundfunk (WDR) studio in Cologne, which made possible the work of composers such as Karlheinz Stockhausen and many others.[26] Moles had worked with the composer Pierre Schaeffer in the early 1950s to develop an early articulation of his concept of the *objet sonore*; with financial support from the Rockefeller Foundation, Moles traveled to MIT to work with Claude Shannon, the chief developer of information theory, and also to Columbia University to work with the composer Vladimir Ussachevsky, who would soon form the Columbia-Princeton Electronic Music Center (CPEMC) with Otto Leuning and Milton Babbitt.[27]

American composers and music theorists interested in burgeoning postwar science and technology quickly took note of these developments. As Eamonn Bell has shown, Moles's work, although it was written in French, was deeply influential not only for European composers such as Iannis Xenakis but also for Americans such as Lejaren Hiller, who drew from his own correspondence with Moles to develop computational music composition software with the ILLIAC computer at the University of Illinois.[28] The composer and theorist Leonard Meyer, similarly, drew deeply from both information theory and cybernetics in his writings on musical aesthetics in the 1950s.[29] Acoustic engineers such as Harry Olson and Herbert Belar, who spent much of the 1950s developing the RCA Synthesizer that would later be housed at the Columbia-Princeton Electronic Music Center, also drew extensively from these discourses. As Eric Drott has shown, the RCA Synthesizer was explicitly designed to be the output stage of a music "analyzer," which would take acoustical sound as input, and through a black boxed internal process, transform it into informational signal that could be resynthesized as sonic output on disc or tape.[30] RCA's imagined system of a music analyzer, encoder, decoder, and synthesizer was explicitly modeled on Claude Shannon's model of a communications channel. Not only could such a system analyze and synthesize music, Olson and Belar speculated; citing Shannon's work on information theory and Wiener's work on cybernetics, they proposed in 1950 that "it does not appear irreverent to music or to the higher facilities of man to suggest a composing machine" that could automate the composition of informationally encoded music itself.[31]

Among these experts, metaphors such as information, feedback, and black boxes were used precisely and specifically in good faith attempts to rigorously translate the sciences of cybernetics and information theory to the domain of music. The promised universality and easy underdetermination of these terms, however, allowed them to be quickly adopted by musicians who had little or no training in scientific disciplines, but who still made efforts to educate themselves in electrical engineering or to work with electrical engineers in order to produce musical spectacularizations of cybernetic concepts.[32] In these cases, as Christopher Haworth has argued, "cybernetics can appear less as a rigorous science and more a kind of practical poetry—an amateur, hobbyist craft concerned with experimentation and art–science cooperation."[33] Haworth, following Eric Drott, has observed that the academic discipline of music studies has largely documented such experiments within "a fairly narrow slice of history (the decades after World War II), a fairly narrow selection of musicians, collaborators, and repertoires (mainly composer-engineer groupings working within the experimental tradition), and a fairly narrow set of pieces that wear the cybernetic influence on their sleeve."[34] Drott has described these cases as exhibiting what he calls the "cybernetic sublime," "in which self-organizing systems and complex feedback mechanisms are staged for public consumption."[35]

Notable examples of musicians embracing the cybernetic sublime in this narrow slice of history and among this narrow demographic include Bebe Barron's modeling of sound-producing electronic circuits after Wiener's *Cybernetics* for her *Forbidden Planet* film score in 1956;[36] David Tudor's development of black-boxed electronics for live electronic music in the mid-1960s;[37] the pervasive thematization of feedback, information, and other cybernetic concepts among the members of the Sonic Arts Group/Sonic Arts Union, including Gordon Mumma's "cybersonic arts," Robert Ashley's feedback pieces, and David Behrman's black-boxed electronics, all also in the mid-1960s;[38] Alvin Lucier's experimentation with biofeedback in his 1965 work *Music for Solo Performer*;[39] and David Rosenboom's developing interest in systems theory and biofeedback in the late 1960s.[40] (The list could go on.)[41] By the 1970s, such cybernetic and information-theoretical concepts, and the technologies that promised to operationalize them in music, were widely distributed around North America and indeed around the globe; by the end of that decade, ensembles such as the League of Automatic Music Composers and composers such as George Lewis used newly available microcomputers to explore networked performance and artificial intelligence, creating what Lewis referred to in an early work's title as "Chamber Music for Humans and Non-Humans" (1979) and later as improvisatory practices with "latter-day musical automata" (1999), which set the stage for the ever-widening fields of computer music, electroacoustic music, electronic music, and research into new human-machine, human-computer, and human-instrument interfaces.[42]

Rather than attempt to account for all cases in which musicians encountered the technoscientific disciplines of cybernetics and information theory and intentionally translated them into the cultural domain of music—which would certainly extend beyond the geographic, social, ideological, and chronological boundaries of the postwar American experimental music tradition and would warrant its own encyclopedic survey—this book attempts to follow a specific thread through its five case studies: the nature of the relationship between the musicking human subject, black-boxed musical instruments, and black-boxed musical thinking. Because cybernetics promised to account for all behavior, both biological and mechanistic, it enabled a radical reassessment of human behavior, as well as the driving force behind it: the human mind. As Bowker has observed, quoting a 1943 article by Wiener and several coauthors, cybernetics allowed for a "uniform behavioralistic analysis" to "both machines and living organisms, regardless of the complexity of the behavior."[43] "As the old dualism between mind and body broke down historically and philosophically," Bowker notes, "features of mind could [...] be found distributed [...] all over nature."[44] Suddenly, everything in nature seemed to be capable of abstraction and purposeful behavior, far beyond the boundaries of the human being.

Through their fantasies about and experiments with novel technologies of electronic sound, in which electricity functioned as a signal understood to be simultaneously cognitive, energetic, informational, vibrational, sonic, and potentially musical, the musicians brought together in this book all had experiences in which their agency as human musicians was called into question by the apparent cybernetic operation of the electronic musical instrument in their hands (or other parts of their bodies). In all cases, to varying degrees, the sound produced by such instruments appeared to be governed, regulated, or organized by some agency outside the human, by the architecture of the instrument itself, or even by the nature of the signal itself. Such black-boxed musical instruments could be schematized as computers that used informational feedback in the service of executing a human's own cybernetic code, but just as easily, they could be schematized as models of the mind itself, bodily prostheses, animals, robots, or models of nature in the most general, abstract, nonscientific sense of the word. At an extreme, such black-boxed musical instruments could even resemble other people. As the composer Suzanne Ciani has reminisced about the "almost human character" of her Buchla 200-series system in the 1970s: "It was a machine of the highest order, a companion, a partner, a friend, and a lover."[45]

BLACK BOXES AND POSTHUMANITY

The consequences that cybernetics and information theory have held for conceptions of the human subject have been extensively theorized and debated among an interdisciplinary group of scholars for more than half a century. One of the most well-known theorizations of these consequences has been N. Katherine Hayles's concept of the "posthuman," in which the human subject becomes essentially informational, fungible between and separable from any given medium (fleshly, electronic, or otherwise) and "seamlessly articulated with intelligent machines." "In the posthuman," Hayles writes, "there are no essential differences or absolute demarcations between bodily existence and computer simulation, cybernetic mechanism and biological organism, robot teleology and human goals."[46] This echoes earlier ontological claims made by cyberneticians such as Wiener that "the physical identity of an individual does not consist of the matter in which it is made [. . .] the individuality of the body is that of a flame rather than that of a stone, of a form rather than of a bit of substance."[47] Hayles, writing fifty years later in 1999, declared that "a defining characteristic of the present cultural moment is the belief that information can circulate unchanged among different material substrates."[48]

As scholars such as Alexander Weheliye have pointed out, however, Hayles's early theorizations of the posthuman ran the risk of flattening the category of the "human" in comparison to the "posthuman" and not accounting for the historically and socially constructed aspects of human identity. Weheliye has argued that "Hayles needs the hegemonic Western conception of humanity as a heuristic category against which to position her theory of posthumanism, in the process recapitulating the ways in which the Western liberal theory of the 'human,' instantiated in the eighteenth century, came to represent 'humanity' sui generis."[49] Thus, to Weheliye, "Hayles reinscribes white masculinity as the (human) point of origin from which to progress to a posthuman state," even as Weheliye acknowledges that other markers of identity, such as gender, play a major part in Hayles's analysis.[50]

To study the ways in which cybernetic and information-theoretical technologies produced varying presentations of humanity and posthumanity outside of the liberal concept of the human, with its unmarked masculine and white identity, Weheliye developed an analysis of raced and gendered human and posthuman vocality in R&B music from the 1970s to the present, focusing on the use of vocoders (a portmanteau of *voice encoders*), instruments that encode sound into quantized signals in discrete frequency bands, originally designed for encrypted military communication. When used by R&B vocalists, Weheliye argues, vocoders, as well as the musical practices undertaken with them, "reticulate the human voice with intelligent machines without assuming that 'information has lost its body' or that any version of black posthumanism must take on an alien form. Because [...] black cultural practices do not have the illusion of disembodiment, they stage *the body* of information and technology as opposed to the lack thereof."[51] Indeed, the question of what constitutes the human, and how the human relates to the machine, was often sonically articulated in Black popular musics throughout the twentieth century. As Kodwo Eshun notes, "there's a heightened awareness in HipHop, fostered through comics and sci fi, of the manufactured, designed and posthuman existence of African-Americans."[52] For Eshun, this posthuman condition is emphasized by instruments such as the Roland 808 Rhythm Composer (which makes possible "the programming of posthuman rhythmatics") and the vocoder (which "turns the voice into a synthesizer").[53]

From Hayles's sweeping conceptual claim to Weheliye and Eshun's responses, theorizations of the posthuman, especially theorizations with material bases in the arts, have often attempted to draw broad historical conclusions—especially from the 1970s onward, as cultural domains such as literature, visual art, film, video art, web art, and video games have more frequently utilized new computing technologies and their related media. As G. Douglas Barrett has suggested, however, the concept of the posthuman is not historically bound to the late twentieth century; it can also be shown to have "ideological predecessors in the human and prehuman."[54] Barrett proposes the concept of the "contemporary posthuman" to refer to the "fractured simultaneity of the pre-/nonhuman, human, and posthuman"; he locates such a contemporary posthuman not in the popular music of the late twentieth century, but rather in a longer arc of experimental music practices, which "can be seen as prefiguring the posthuman in the ways indeterminacy seeks to remove the human from the creative process."[55] Here Barrett specifically refers to the practices of indeterminacy developed by John Cage, which he argues were central to the development of both American and European experimental music communities in the 1950s and beyond. For Barrett, "postwar experimental music composes the contemporary posthuman."[56]

PLAYING WITH BLACK BOXES

While the musicians brought together in *The Composer's Black Box* could be grouped into the category of "postwar experimental music," and while their work has much to say about conceptualizations of the posthuman, this book does not attempt to make a singular theoretical claim about the posthuman condition or a singular claim about postwar experimental music. Indeed, this book did not begin as a theoretical study of posthumanism or as a historically bounded monograph about experimentalism in music at all. Rather, it emerged from a question I often found myself asking while I had my hands on the electronic musical instruments that emerged from the postwar American experimental tradition that, qua Hayles and Eshun, as well as their manufacturers' marketing copy, promised to blur or even eliminate the "demarcations between bodily existence and computer simulation, cybernetic mechanism and biological organism, robot teleology and human goals" (Hayles), in turn producing a "hyperembodiment" that promised to make me "feel more intensely, along a broader band of emotional spectra than ever before in the 20th century" (Eshun).[57] As a thoroughly posthuman subject in the 2010s, both of those conditions sounded intriguing to me. The question I had, as I started to experiment with these instruments, was: Why was it so hard to figure out how these instruments worked and to get them to produce the sounds that would make me variously lose and/or regain my sense of self? Why was I not feeling the blurred lines between myself and the instrument, or a sense of hyperembodiment? Why wasn't I becoming posthuman?

My immediate thought was that I simply needed to *learn* to be posthuman, to spend more time with the instruments. But the more time I spent with the instruments, the less I felt like a posthuman, and the more I felt like something much more ambiguous. The particular instrument I was using was a Buchla Music Easel (a specific configuration of two of his 200 series modules), which I had purchased in 2016 with money from an entire semester's worth of my graduate stipend from the University of Chicago. Its dizzying array of color-coded jacks, sliders, and switches did not provide me with an easy way in; its user manual, titled *Programming and Metaprogramming in the Electro-Organism*, was even more opaque.[58] It took me hours to figure out how to even produce a sound with it. I came to this Buchla-designed instrument through my dissertation research about the SFTMC, a storied institution that helped to launch the careers of composers and technologists such as Morton Subotnick, Pauline Oliveros, and Ramon Sender, and which facilitated the development of what I then believed to the first voltage-controlled modular synthesizer—the Buchla Box. So much ink had been spilled about the importance of this system, its heterodox design, and the unique possibilities it afforded, especially in comparison to the more commercially successful Moog synthesizer, that I knew I needed to get my hands on a Buchla instrument in order to fully understand it. I had read Subotnick's description of a music easel that would allow a human to paint music and had listened to his albums of ostensible musical painting such as *Silver Apples of the Moon*; I had even spoken with him directly about his utopian technological narrative, in which he remembered speculating that every home would have a Buchla Box that would externalize its owners' innermost creative ideas without the burdensome expense of conservatory training. So, when the Easel arrived, I gleefully opened the briefcase in which it was housed and attempted to do just that.

No musical painting was done that day. Using a Buchla instrument, I initially thought, seemed to be a textbook example of a canonic concept I had been reading about in a science and technology studies seminar: Andrew Pickering's concept of the "dance of agency," a "dialectic of resistance and accommodation" between a scientist and their instruments that forced the scientist to revise their intentions and also the "material form of the machine in question" in order to produce scientific knowledge.[59] But the Buchla Box was not an instrument of science; it did not force me to revise a research question in the name of producing epistemological truth. And not only did this Buchla instrument show me that I could not externalize my internal musical ideas; it also made me question what those internal musical ideas were in the first place—*whose* ideas they were—and how they emerged through this dance between my mind, my body, and my instrument.

BLACK-BOXED MUSICALITY

Despite the questions this instrument forced me to ask about the internal workings of my mind and body—what I wanted, what I could sense, what made me "creative," what "music" even was or could be, and how all of those questions could be articulated with and through the instrument in front of me—I found that other musicians often avoided discussing the internalities of the relationships between themselves and their electronic musical instruments, and instead preferred to discuss the externalities of such relationships, that is, the sounds they produced. For decades, Subotnick towed the line about his composer's black box enabling his concept of "music as a studio art," sitting for countless interviews in which he rearticulated his utopian vision from the 1960s while almost religiously avoiding discussing the labor-intensive and often frustrating processes of working with his instruments. It wasn't until 2024 that Subotnick admitted he was "terribly disappointed with all of the modules [on the first Buchla system] [. . .] we had made terrible mistakes."[60] Those mistakes were perhaps why it took him thirteen months of working with the instrument to produce *Silver Apples of the Moon*, a far cry from the immediacy of painting sound. But these "mistakes," to me, seemed to be features, not bugs, in the "analog computer for music" that comprised the composer's black box: the inscrutability of Buchla's modules, their wildly unpredictable and nonlinear behavior, and their tendency toward complexity and opacity rather than simplicity and transparency when patched together to form an instrumental system.

This was my first clue that black-boxed musicality was not exactly the same as posthumanity. Subotnick *wanted* to seamlessly articulate himself through his Buchla Box, but it never quite worked like that. Initially I thought that this black-boxed behavior was limited to the Buchla Box, since Buchla was so steeped in the cyberculture of the 1960s, and since other instruments like the Moog had musical keyboards that might allow for more immediate articulation. In the Buchla, nothing was hardwired; the system did not have a musical keyboard; every time a human used it, a new configuration would need to be explored and auditioned. But rather quickly I came to realize that this mode of black-boxed musicality was present even with musicians and instruments that could appear to be the opposite of Subotnick and his Buchla Box.

When I first heard Sun Ra's 1970 performances with his prototype Minimoog—an instrument with hardwired connections between its electronic controls and a musical keyboard that controlled the fundamental frequency of its oscillators—I heard inscrutable, wild, unpredictable musical performances that stood in contrast to Ra's more controlled performances of his compositions for his Arkestra and even his more intense improvisations with electric organs and other keyboard instruments. And when I read what Ra wrote about those performances, that instead of expressing human freedom they demonstrated the precision and discipline of a more-than-human, cosmological authority, I knew that the opacity of black-boxed musicality was not limited to Buchla. For Ra, the Minimoog served as a black-boxed model for his own black-boxed subjectivity: opaque, yet following an unknowable internal agency that Ra located outside the human, somewhere out there in the cosmos.

By the mid-1970s, both Subotnick and Ra had abandoned their early black-boxed instruments in favor of ones that appeared to offer more easy and reliable human control.[61] But in the crucial moments of their first encounters with electronic musical instruments—instruments that used electricity to communicate information, not transfer energy—both produced ambiguous, emergent musical performances with musical black boxes, instruments that opacified the music-making process between human and instrument, making possible new understandings of each in turn. The difference between the two was that each drew different consequences for understanding themselves in relation to their instrument. Subotnick, faced with his composer's black box, chased a techno-utopian dream of the perfect expression of the bounded, liberal self. Ra, faced with his prototype Minimoog, developed his own techno-utopian dream in which the self was unbound to planet Earth, following its own extraterrestrial, more-than-human logic.

I began to wonder about other encounters that musicians had with black-boxed electronic musical instruments in the 1960s, before the 1970s saw the standardization and commercialization of the "synthesizer" as a keyboard instrument that could simply produce new sonic timbres. Unlike Subotnick and Ra, there were many other musicians whose encounters with musical black boxes in the 1960s (and even as early as the 1950s) were not with systems intentionally designed to be musical, but rather with other electronic technologies that used electricity to convey sound as information: oscillators, tape machines, contact microphones, and a wide variety of electronic audio technologies that were becoming increasingly smaller, cheaper, and more available outside their intended markets. I've already listed examples of composers who explored what Drott has called the "cybernetic sublime," mostly associated within a narrowly defined swath of European and American institutions. But examples also abounded outside of the academy, including Les Paul's famous experiments with "sound on sound" tape recording (1948); Raymond Scott's staggeringly complex "Electronium" (1959); Rahsaan Roland Kirk's integration of electronic devices and tape machines in his multi-instrumental setup (1965); and Simeon Coxe III's eponymous, home-brewed assemblage of oscillators, switches, and telegraph equipment used in his band Silver Apples (1968), to name only a very few.[62] These figures and instruments all certainly warrant their own study, especially Kirk, as one of the earliest adopters of electronics in jazz.[63] In this book, however, I chose to focus on cases in which the challenge of black-boxed musicality was most intense, the stakes most high, because it threatened the social identity of "the composer" in ways that could, and did, fundamentally change that social identity, leading to a variety of pathways through musical life in the second half of the twentieth century.

Between the two poles of Subotnick and Ra, *The Composer's Black Box* explores three other cases in which the human subject was more ambiguously articulated in relation to musical black boxes—in which the black-boxed musicality that emerged between human and instrument produced differing consequences for the ontology of the human and nonhuman (or more-than-human, other-than-human) agents involved in making music. In contrast to Subotnick's early fantasies about the Buchla Box being a composer's black box, chapter 2 examines how Buchla's systems were used by amateur musicians and psychonauts in the 1960s San Francisco counterculture, developing a computational understanding of the human mind that allowed for ecstatic psychedelic expansion, both electronic and chemical. Chapter 3 then shifts to the composer Pauline Oliveros's early electronic systems, also constructed at the SFTMC, that enmeshed her mind and body in electronic-sensory networks of oscillators, tape delays, and loudspeakers, and explores the consequences that these systems had for Oliveros's understanding of her own body as a queer woman. To further explore the technicity of the human body and its socially constructed markers of identity, chapter 4 turns to the contemporaneous early electronic work of the composer Alvin Lucier, whose use of electroencephalograms, contact microphones, and various parascientific audio devices suggested visions of hybridized, cybernetic beings between human, animal, and computer in a playful and ambiguously poetic register.

In all of these cases, musicians encountered the technologies and metaphors of cybernetics and information theory and through them developed musical practices that differentially (re)imagined the musicking human subject in relation to their electronic instruments and the signal that flowed between them. While the emergent, cybernetic musicking subjectivities developed by some people in this book were conceptualized within an axis of "human/nonhuman," or "human/posthuman," which ran the risk of reinscribing the implicit and unmarked identities of white masculinity onto the category of the human, others swerved from these unmarked identifiers by explicitly imagining the human through other schema of identity, including gender, race, and ability. The stakes of the five case studies brought together in this book were no less than the future of the human subject in cybernetic America, engaged in an activity that questioned what had long been imagined to be a profoundly stable marker of humanity: making music.

A NOTE ON SOURCES

The 1960s were a long time ago, and to quote a famous dictum, "if you can remember the sixties, you weren't really there."[64] The encounters that the musicians in this book had with technologies in that decade (and often as far back as the 1950s) became the subjects of anecdotes and life stories that were told and retold for decades thereafter. Yet those encounters often left no trail of paper or audiotape; in many cases, diaries and letters were lost, tapes were lost or recorded over, and nobody thought to document what seemed like ongoing processes that would eventually, for many of the people studied in this book, produce compositional works—the objects that mattered to the institutions that paid their bills. By the 1970s, however, it became clear that these early moments in the 1960s were important, and what had been hazy memories were quickly shored up, anecdotes were concretized, and dates came to be attached to now canonized moments in order to patch together coherent historical narratives.

This book is full of memories, and it is also full of archival traces—and aporias. In many cases, memories and what does or does not remain of a paper trail are incommensurate, producing ambiguities that are consequential to the artistic, social, and economic stakes of these musicians' legacies and ongoing careers. (Ra passed in 1993, though the Arkestra lives on under the leadership of Marshall Allen, who turned 101 in 2025. Throughout the course of writing this book, Buchla, Oliveros, and Lucier have passed; Subotnick remains.) In several instances, I have chosen to write about memory as fact in the body of this book and to provide a more complete explanation of what archival traces do and don't exist in the book's notes. Thus, for example, chapter 2 begins with Subotnick's memory of placing a classified advertisement in the *San Francisco Chronicle* for an engineer to build his composer's black box sometime between 1962 and

1964, even though no such specific advertisement can be found in archived copies of the *Chronicle* from that period. (It may be that I am simply a bad historian, but many days of poring over microfilm and searching digital databases turned up bupkis.) Despite the occasional absence of a paper trail or the presence of conflicting information, I find these memories important enough to include in the body of this book's narrative because they form the basis of these musicians' social identities. And it is these identities that I am most interested in as they emerge from, pass through, and become transformed by encounters with science and technology—new ways of knowing, being, and doing in an increasingly technoscientific world.

1 Morton Subotnick and the Composer's Black Box

A screaming comes across the speakers: an almost pure sine wave with a few odd harmonics, periodically jumping in fundamental frequency between the narrow band of 1,600–1,950 Hz, punctuated by millisecond bursts of higher frequency sawtooth waves between 2,600–6,000 Hz, before the introduction of lower, frequency-modulated sine waves, all while hisses of white noise flood the audible spectrum in brief irruptions. Within seconds, multiple discrete parameters of this screaming begin to be modified—frequency, amplitude, attack, duration, periodicity, rate of change—until it abruptly fades, making space for a new sound, a gradually undulating sawtooth wave around 175 hertz, perceived as a low F, which heralds an even faster barrage of filtered white noise bursts.

These are the opening moments of Morton Subotnick's 1967 LP *Silver Apples of the Moon*. If you didn't understand this description of it, don't worry; an average listener in 1967 would not have, either. This music is not shaped by familiar parameters of pitch, rhythm, harmony, or melody, but rather those of frequency, waveshape, modulation, attack, duration, and periodicity. Perhaps anticipating the difficulty for the audience to understand the musical content of the album, the album's liner notes instead focus on its historicity and technology, claiming that "this album of electronic music represents a signal event in the related history of music and the phonograph: for the first time, an original, full-scale composition has been created expressly for the record medium."[1] A well-informed reader in 1967 might have recognized the term *medium* from Marshall McLuhan's 1964 book *Understanding Media*, which spawned the popularity of the phrase "the medium is the message." The album's claim to specificity to the "record medium," perhaps, *was* its message: "The idea of writing a work especially for recording presents the composer with a special frame of reference.... [I]t is not the reproduction of a work originally intended for the concert hall.... [R]ather it is intended to be experienced by individuals or small groups of people listening in intimate surroundings ... as a kind of chamber music 20th-century style."[2]

If the work on this record was chamber music, what made it "20th-century style"? Not only was it specific to a twentieth-century medium, but it was also, as the album's notes remind the listener, created with an "electronic music synthesizer," a definitively twentieth-century technology. The synthesizer used to produce *Silver Apples of the Moon*, however, was not a musical instrument like a violin or a piano, with which a human's physical gesture would reliably produce a discrete, repeatable sound. Instead, Subotnick characterized this instrument as

> an electronic music "machine" that would satisfy our needs as composers. The system generates sound and time configurations, which are predetermined by the composer through a series of "patches" consisting of interconnecting various voltage-control devices. It is possible to produce a specific predetermined sound event... and it is also possible to produce sound events that are predetermined only in generalities.... [T]his means you can "tell" the machine what kind of event you want without deciding on the specific details of the event... and listen... and then make final decisions as to the details of the musical gesture. This gives the flexibility to score sections of the piece in the traditional sense... and to mold other sections (from graphic and verbal notes) like a piece of sculpture.[3]

It is relatively clear what it might have meant to "produce a specific predetermined sound event." But what did it mean for a machine to be a system of interconnected modular devices, rather than a singular instrument? What did it mean to "tell" that system to produce a sound event that is "predetermined only in generalities"? What were the different possible "kinds" of events? In the absence of the physical material of magnetic audiotape, what exactly was being "molded" like a piece of sculpture? Who—or what—did the sculpting?

Examining the "machine" with which Subotnick created this album does not exactly answer these questions, but rather clarifies why they need to be asked. The Modular Electronic Music System, designed by Donald Buchla and commonly called the Buchla Box, was a system of interconnected modules that behaved both like an analog computer and like a cybernetic, self-governing system. Because it executed "programs" patched by its human user through its various jacks and knobs on its front panel, it behaved like an analog electronic computer. But insofar as those programs almost always by necessity included elements of feedback, automatic control, and real-time self-modification, that computer was also cybernetic. The distribution of control within this system challenged unilateral notions of human composerly agency; once activated, the system essentially played itself, often in ways that its human user could not anticipate, and indeed could not control. This challenge meant that producing *Silver Apples of the Moon* was not as simple as composing a program for the system, executing that program, and recording the result to tape. Instead, by Subotnick's estimate, it took thirteen months of working eight to ten hours per day in a feedback loop of programming, listening, adjusting parameters, listening again, and recording the output of this interaction to tape—eventually recording enough tape to overdub and edit sections together to produce the work as a whole.

Although Subotnick edited tapes together to produce the work, the sounds on those tapes emerged from a human-instrument interaction that challenged the position of the human as the ultimate origin of the organization of sounds in time. Because of the Box's instability, nonlinearity, and cybernetic systematicity, it potentially shifted the locus of musical agency away from the human and toward the system itself, allowing for new, unexpected sounds and organizations of sounds to emerge from the iterative, systematic interaction between human and instrument. Indeed, as discussed in chapter 2, many of the Buchla Box's earliest users reveled in the promise of such a cybernetic subjectivity, especially since it mirrored what many of them believed to be the psychedelic expansion of the mind—new sensory and cognitive modalities made possible through technologies both chemical and electronic. But for Subotnick, such cybernetic subjectivity, and the threat it posed to the notion of traditional composerly subjectivity, was a challenge to be not embraced, but rather overcome.

If the cybernetic subjectivity of the Buchla Box posed such a challenge to the agency of the composer, then why did Subotnick desire the creation of such a box in the first place? The Buchla Box emerged from a period from the late 1950s into the mid-1960s during which Subotnick developed a totalizing concept of musical composition, "music as a studio art," in which a single human being would express the deepest core of their individual subjectivity by creating not only the score for a musical work, but also every other element of such a work from the bottom up, beginning with the very instrument on which it was conceptualized. As Subotnick remembered in 2017, "I was looking for a new kind of music that you could do where you were in charge of everything. An instrument maker, performer, listener and conductor. That's what music as a studio art means."[4] To create such music as a studio art, Subotnick envisioned a "music easel," a neutral, invisible, architectural support for the canvas of sound; and to operationalize such an abstract concept, Subotnick imagined what he later called a composer's black box, an "analog computer for music."[5] Borrowing concepts from the contemporaneous technoscientific disciplines of information theory and cybernetics, such a device would be not only an architectural system through which each individual user could specify the parameters of their own unique musical instrument, but also a device that would remain simplified into input and output—"black-boxed"—so that the user would not need to understand how it worked. Such a box, Subotnick imagined, would allow anyone, even without any musical training, to "paint" music intuitively, gesturally, and individually.

Silver Apples of the Moon has been widely hailed as the first example of such music as a studio art, and the Buchla Box as the first instantiation of Subotnick's composer's black box. But rather than read this work as the origin of the concept, in this chapter I read it as the end of a years-long trajectory that led Subotnick to envision the ultimate expression of the autonomous, individual liberal subject in the first place. By following Subotnick's early exposures to the technoscientific social and ideological world of American New Music composition in the late 1950s, his earliest experience with tape recorders and other electronic media, his involvement with the black-boxed technoaesthetics of experimental theater, and his encounter with and novel interpretation of the media theory of Marshall McLuhan, this chapter shows how, and why, Subotnick imagined the composer's black box to be the ultimate extension of his creative self. For Subotnick, the metaphor of the black box promised a simplification of technological processes that would allow for an easier and more fluid expression of individual subjectivity. This marks the beginning of an arc that I trace throughout all five chapters of this book: from the black box as a site of the perfect articulation of the liberal human subject to the black box as a site of that subject's profound opacity.

BECOMING A COMPOSER: STOICISM, LIBERALISM, AND AMERICAN TECHNOSCIENCE

In a July 1964 radio interview broadcast from the University of Cincinnati College-Conservatory of Music, Subotnick, then thirty-one years old, was asked to summarize his work as a composer. Describing the previous six years of his public career, Subotnick mused, "My [work] has a reference point in a kind of a broad range of 19th century mythology, in a way, if there is such a thing, and it's something I've been for the last several years extremely involved with."[6] Subotnick formulated this response to explicitly contrast himself to his colleague, Ramon Sender, with whom he had cofounded the SFTMC in 1962. Sender, whom Subotnick described as inhabiting a "20th-century mythology," wanted to do away with the concept of the individual artist—which he derided as a "masterpiece machine"—and to instead move toward a more communal modality of creating work, even toying with the idea of completely anonymous concerts with no composers credited.[7] Both composers had been working with the new technologies of magnetic audiotape and various electronic sound-producing devices for nearly six years. But where Sender saw in these new technologies an opportunity for a new pluralistic modality of community-oriented creativity, Subotnick saw in them a golden opportunity for the individual:

> Once a composer starts using tape, and starts using electronic and other means in tape, there are no restrictions, there are no boundaries, and there

is no precedent. So that many of the issues that you grow up with as a music student, composition student—many of the problems that you have been faced with in terms of the problems of tonality and the problems of atonality, problems with form and structure and content and so forth, absolutely don't exist. Not that the problems aren't there. But there isn't a Beethoven of the tapes to refer back to. And so you rely entirely on your own imagination, and what it does is really pique the imagination.[8]

Subotnick's response implies two nested propositions about what it meant to be a composer in the 1960s. The first is that a composer is an individual human subject who should be free to express their imagination through music—reminiscent of Subotnick's "19th-century mythology" of music as the ultimate expression of that's subject's profound interiority. The second is that the technologies of tape and electronic instruments emancipate the human subject's imagination from sonic restrictions, formal boundaries, and historical precedents, further purifying music into the realm of the individual imagination.

The first proposition articulated a long-held ideological belief that Subotnick has traced back to a childhood encounter with the Roman emperor Marcus Aurelius's text *Meditations*, found in a "great books" set in his parents' garage, which contained aphoristic summaries of the major Greek Stoic philosophers. Subotnick summarized this belief as an ethical imperative: "The meaning in life is to discover who you are and what you're capable of, and to do it the best you can, and to share it with other human beings. [. . .] That was the decision. I didn't know who I was or what I had to share, but I knew that that was what you had to do."[9] Although Aurelius's writings instruct the reader to undertake the philosophical exercise of *askēsis*, or self-discipline, such practice is limited to the self. Subotnick's imperative to "share [the self] with other human beings" seems to be his way of connecting Aurelius's Stoicism with the social activity of musical composition; Stoic *askēsis* was similar to the disciplined introspection that would produce a musical composition, which would then be shared with the world through its performance.

But as Subotnick complained in this 1964 radio interview, his self-discovery through musical composition had been hampered by restrictions, boundaries, and precedent. Up until 1959, Subotnick had been struggling as a composer, creating works in a post-Webernian atonal style that did not seem to satisfy his teachers Leon Kirchner or Darius Milhaud, although both were extremely supportive of his work. These early works did not seem to satisfy audiences, either: at one of the first major public performances of one of Subotnick's compositions, a four-hand piano sonata at the Aspen Music Festival in the summer of 1959, the audience talked through most of the work's performance and started filing out during the last movement. Subotnick was despondent, despite Milhaud's delight at the scandal.[10]

Subotnick's early struggle with composing for acoustic instruments apparently motivated him to turn to what his colleague Nate Rubin called "a newly discovered electronic universe."[11] Subotnick's initial exposure to this universe occurred immediately after the Aspen Festival in August 1959 at the Princeton Seminar in Advanced Musical Studies, a program established by the wine merchant and music philanthropist Paul Fromm to serve as an American answer to European seminars such as the Darmstadt Summer Course. The seminar was the first successful instantiation of several years of work to establish some kind of institute for new music at Princeton on behalf of the composers Milton Babbitt, Roger Sessions, and Ernst Krenek; at various points, both Krenek and Babbitt enlisted support for such an institute from Henry D. Smythe and Robert Oppenheimer, both of the Manhattan Project.[12] World historical in ambition, the seminar prided itself on its prestige: "Designed for study, on the highest level, of the directions followed by the most advanced musical thinking of the present day, the seminar is intended to provide an answer to what its sponsors feel to be a genuine need in the musical life of both the United States and the world at large."[13] Subotnick was invited both as a composer and as a clarinetist, performing Babbitt's *Composition for Four Instruments* (1947–1948) in addition to participating in seminar meetings.

Babbitt's leadership role at the seminar took place during a crucial point in his career when he was formulating his own ideological position on what it meant to be a composer in Cold War America, acutely shaped by his recent work with the RCA Mk. II Sound Synthesizer, which he first encountered in 1957 at RCA's Sarnoff Research Center in Princeton, New Jersey, and which had subsequently been installed in the CPEMC. For Babbitt, working with the RCA Synthesizer helped to articulate a political ideology about the experimental composition of new music that, as Martin

Brody has argued, "supported a vision of human rather than mechanical potential—and of the autonomous artist as a Cold War liberal, a heroic figure more enduring than RCA's celebrated machine."[14] In other words, although the machine enabled the precise specification and production of various parameters of sound, for Babbitt such a machine ultimately served to emancipate the individual human composer by enabling them to liberally use their imagination to its fullest potential, without historical precedent.

In a 1961 article, Babbitt summarized this position: "Present-day electronic media for the total production of sound, providing precise measurability and specifiability of frequency, intensity, spectrum, envelope, duration, and mode of succession, remove completely those musical limits imposed by the physical limitations of the performer and conventional musical instruments. The region of limitation is now located entirely in the human perceptual and conceptual apparatus."[15] Although Babbitt still speculated about "limitations" on an individual human's "perceptual and cognitive apparatus"—as Brody points out, a view strikingly similar to that of the psychologist and eugenicist Carl Seashore as well as his student Harry Olson, who designed the RCA Synthesizer—these limitations were only relevant insofar as they could help distinguish between universal limitations of a human brain's "memorative capacity" and "the discrimination between aural quantizations imposed by physiological conditions," and "those imposed merely by the conditioning of prior perceptions," that is, enculturation to "common practice" musicality.[16] In addition, to Babbitt, those "perceptual and conceptual capacities" were not static, inheritable traits, but were rather dynamic capacities of an individual human being, subject to "explosive and decisive extensions" through recent technoscientific developments. As Brody summarizes, electronic music technologies such as the RCA Synthesizer "sparked a revolution in musical thought that was unending, but circumscribed by liberal values—principles of pluralism, critical speech, and intersubjective communication."[17]

Babbitt's liberal ideology of American New Music shaped the contours of Subotnick's experience at the Princeton Seminar in 1959; that ideology was operationalized through the technoscience of information theory. Three afternoons per week of the three-week seminar were dedicated to lectures on scientific theory and musical theory, which covered logical propositions and group theory, electronic music, and information theory, respectively.[18] In addition, the mathematician John Tukey also visited the seminar to discuss information theory; Tukey, a member of Princeton's Mathematics Department, had advocated for Babbitt's PhD dissertation to be accepted after it had been rejected by the Music Department, and had also worked directly with Claude Shannon and John von Neumann, two foundational theorists of information and systems theory.[19] During the second week of the seminar, Babbitt addressed "the history of the electronic medium, the physical properties of a tone, spectrum analysis, additive and subtractive synthesis, formants, envelopes, and the properties and layout of the RCA Mk. II Synthesizer."[20] Seminar attendee Carlton Gamer recalls that the seminar participants traveled to visit the RCA Synthesizer in person, with Babbitt demonstrating his "elaborate preparatory setup" of the instrument (which inadvertently turned itself off because of an error in its programmed instructions).[21]

Thus in August 1959, Subotnick was introduced to both an abstract, scientific paradigm of music and an elaborate technological instrument that promised to operationalize that paradigm. For Babbitt, who served as Subotnick's guide to this brave new world, such a paradigm promised that "any perceptually meaningful musical event can be specified and obtained," which meant that the only constraint on the creation of music lay in the individual's "conceptual apparatus."[22] Such electronic instruments were presented to Subotnick as agnostic to history, style, or pedigree; they reaffirmed the fantasy that each user could specify, and ultimately produce, musical events that were wholly unique to their own individual artistic imagination. For this reason, as Brody has argued, the Synthesizer was for Babbitt a "potent enabler of liberal imagination," representing a particularly American sensibility of "cosmopolitanism and epistemological acuity" that "overturned the metaphysics of musical genius in the name of epistemological/artistic pluralism."[23] Brody notes that Babbitt's embrace of information-theoretical technologies such as the RCA Synthesizer "was aimed not at defining a common musical practice, but rather exercising a communal mode of inquiry [...] Experiencing the collisions of autonomous *systems* in the pluralistic world of contemporary music might fortify the autonomy of liberal *subjects*."[24]

Although Subotnick did not have the opportunity for further work with the RCA Synthesizer, immediately after the Princeton Seminar in the fall of 1959, he began to put this liberal ideology into practice when he found his first artistic and financial opportunity to work with a tape recorder, the most basic of electronic instruments, which he remembers as emblematic of "the new technology rising on the music horizon."[25] Through his former teacher Leon Kirchner, Subotnick had been introduced to Herbert Blau, the director of the Actor's Workshop of San Francisco, who was looking for a composer to create music for a new production of Shakespeare's *King Lear*. The Actor's Workshop was a theater company that shared Babbitt's ideological commitment to democratic liberalism and world-historical artistic work; despite some members being blacklisted by the House Un-American Activities Committee (HUAC), its production of *Waiting for Godot* was selected by the US State Department to be performed in the American Pavilion at Expo '58 in Brussels.[26] And for its new production of *King Lear*, Blau sought a "wild young composer" to help him achieve his political goals, using the theater to "express the whole rhythm of social intelligence—the most full-bodied treatment, more immediately human than that of architecture, to men in an active *believing community*, moving by right reason and concerted instinct toward some valid if barely-realized purpose."[27] With an advance of $150, Subotnick purchased a portable two-track Grundig tape recorder and got to work.

ENCLOSURES AND ATMOSPHERES: *KING LEAR* AND *SOUND BLOCKS*

In Subotnick's earliest work with new electronic audio technologies—his score for *King Lear* and a subsequent related work, *Sound Blocks*, discussed later—sound was used to enclose the audience both physically and metaphorically, ostensibly sharing the composer's imagination with the audience in a totalizing theatrical experience. As Subotnick remembered of his work for *Lear*, he aimed to "create an audio score that would not be incidental music but like the set, lighting, and costumes, would be integrated into the physicality of the action to inform the unfolding of the drama."[28] Subotnick's score centered around the "storm scene" in *Lear*, in which the king famously declares: "When the mind's free, / the body's delicate: this tempest in my mind / doth from my senses take all feeling else / save what beats there."[29] Lear suggests that the tempest within his own mind—what "beats" there—creates its own sensorium, allowing him to ignore the physical sensations of his body caught in the storm. With his tape machine and loudspeakers, Subotnick sought to collapse the distinction between these two registers, suggesting that inner and outer experience could be integrated into a single, totalizing sensory experience in the enclosed space of the theater.

Blau remembered Subotnick's work as creating a totalizing "atmosphere" that comprised "the sound of Lear's voice saying the word 'I' into an open piano; a single pure pitch; and a cello note," and that "imposed on this was another sound track of accidental electronic sounds."[30] These sounds produced what Blau described as "space," a "steady presence" of the storm that would be "integrated into the physical presence of the action." Blau described Subotnick's score as an agent somewhere between actor and mise-en-scène, allowing the two to bleed into each other: "The storm was *music*, with its own properties of perception, and its own interior life [. . .] Artaud speaks of the *mise-en-scè*ne as actor. What we wanted was for the actor to create the *mise-en-scè*ne as it was creating him. The atmosphere dictated the possibilities; the actors, struggling against the storm, became it."[31]

Following the idea that "the atmosphere dictated the possibilities," Subotnick began to develop a new work entitled *Sound Blocks: An Heroic Vision*, which would physically enclose its audience in an atmosphere created through light, sound, and scenic elements. For the 1961 premiere of this work, Subotnick assembled four lighting rigs made from aniline-dyed cotton scrims with attached abstract sculptural objects that would change dramatically in appearance depending on their lighting, and positioned them in a box-like formation around both audience and performers.[32] As the critic Alfred Frankenstein recounted, a musician would emerge alone from the corners of this box into the middle of the space, and "a piercing spotlight picked him out of the almost totally darkened room and remained on him for the duration of his performance."[33] Each of the four musicians performed their acoustic score along to the sound of prerecorded tape, which Frankenstein described as "palpably textured, like wool or rough metal."[34]

While the musicians played, a poet recited aloud Michael McClure's poem "The Flowers of Politics," which celebrated the discovery, reinvention, and celebration of the self through a renewed consciousness of the human body's sensory relationship to its environment.

The poem begins:

THERE IS THE HUGE DREAM OF US THAT WE
ARE HEROS THAT THERE IS COURAGE
in our blood
That we are live!
That we do not perpetrate the lie of vision
forced upon ourselves
by ourselves. That we have made
the nets of vision real!
AND SNARED THEM[35]

And ends:

This is
A CANDLE
and shape of light.
The hand and arm annunciate all things
and draw the eye upon the speaking face
declaring from the inner body. We are wrought on a bending shaft of air and light
and make an animal around it
and spread a radiance from ourselves that melts in light.[36]

McClure's poem exhorts the reader to deny the "lie of vision / forced upon ourselves / by ourselves" and to reassess their sensory relationship to the phenomenal world to "make an animal" around a "bending shaft of air and light." The reader is encouraged to first appeal to the senses and then, using the ephemeral material perceived by those senses—light, in this case—to remake the self. In the case of *Sound Blocks*, this was quite literal. The audience was surrounded by a novel phenomenal "atmosphere" that produced novel sensory phenomena that, to Subotnick, productively alienated them from their sense perceptions in "heroic" and "courageous" ways.[37] As Subotnick remembers it, the response of the audience and the press was extremely favorable: "It was a sensation. I mean people were wiped out. We got offers to keep doing this. We performed every Sunday night for three weeks, or something like that. Reviews in the newspaper were saying a new art form had been born."[38]

Indeed, Frankenstein's review called *Sound Blocks* a "masterpiece." But perhaps more importantly, Frankenstein picked up on Subotnick's "environmental" metaphors, describing the "new universe of rhythms, tonal colors, and tonal textures brought into being by the tape recorder" and fantasizing that "the live sound [of instruments] was of the world of ordinary experience, and the recorded of some streaming, busy, fantastic world." Moreover, Frankenstein described the work as producing a sense of enclosure: "The whole resonating surface of the room—walls, floor, and ceiling—became a musical instrument, and when everybody was at work in the shattering climax of this 'heroic vision', one understood the sentiments of a mosquito caught inside a violin while Heifetz is playing the Brahms concerto."[39] Within the enclosure of this performance of *Sound Blocks*, Frankenstein imagined being inside the instrument of a European virtuoso playing a masterpiece by a great man. But for subsequent performances of this work, the valence of such an enclosure would change dramatically, reflecting a more concrete engagement with contemporary political and technoscientific trends. Not only would the space be enclosed; it would become a literal "black box."

For the second performance of *Sound Blocks*, installed in a theater lobby during the Actor's Workshop's 1962 revival of *Waiting for Godot*, Blau envisioned the theatrical space as the site of an underground nuclear test—not only a site of sublime terror, but also a site of sublime possibility, with jewels emerging from the release of nuclear radiation: "I happened to read an account of an underground nuclear test, in which the released megatons accelerated the process of mineral evolution, so that artificial jewels were embedded in the ground. It was such a landscape that I had in mind for the new production [of *Godot*]. The audience would be in the cave (our little theater has a low ceiling); the stage would look as if it were blown away from the end of the building by a blast. The whole landscape would look *man-made*."[40]

Although this production predated the first published use of the term *black box* to describe a theatrical space by nine years, it was quite literally black: the floor of the stage was covered with foam rubber layered with black latex, which Blau likened to Artaud's concept of *bubos*, the abject secretions in the theater of cruelty that appear "wherever the organism discharges either its internal rottenness, or, according to the case, its life."[41] Blau also remembers this production of *Waiting for Godot* as the first time they began to experiment with removing the front curtains of the proscenium stage, attempting to make the action take place in a frame that, like an abstract expressionist painting, widened the canvas in order to avoid interaction with its formal edges. Blau called this "working inside the box to exceed its limits": an attempt to fully inhabit the black box without acknowledging the boundaries that separated it from the rest of the world.

With *Sound Blocks* installed in the lobby, Blau prepared the audience for his black box with another experience of enclosure: "This action and the blitzed landscape [of *Waiting for Godot*] were prepared for by a score of *Sound Blocks* by Morton Subotnick, electronic music composed directly on tape. The audience walked into this barrier of sound, this ambiance, as they entered the auditorium. They could not adapt to it because of its atonality and the accidental occurrence of its sequences, until, like a disarticulated Pied Piper, it led them directly into the play."[42]

In Blau's memory, *Sound Blocks* created a "barrier" that prepared the audience to enter the nuclear test site within the theater, wherein they would experience what the anthropologist Joseph Masco has called an aestheticized and intellectualized form of the "nuclear sublime" that no longer posed a "visceral threat to the body of the scientist."[43] In Blau's directorial vision, the terrible power unleashed by contemporaneous American technoscience created artificial jewels and new aesthetic forms of beauty among the rubble. Indeed, as Masco has argued, the technoaesthetics of nuclear tests in the 1950s "provide[d] the U.S. public with an image of the bomb as transcendent form, minimizing the physical effects of the explosion in favor of the conceptual power of a 'new age.'"[44] That conceptual power of a new age, of better living through technology, would continue to motivate Subotnick even as he moved on from *Sound Blocks* to begin other projects that would more explicitly embody the technoaesthetics of such a new era, namely through another major contemporaneous technology: the computer.

THE COMPUTER AND THE MIND OF MAN

In 1962, one year into working with electronics, Subotnick was commissioned to create a score for a television special entitled *The Computer and the Mind of Man*, a six-episode series produced by the National Educational Television and Radio Center and aired locally on KQED-TV, which featured interviews with researchers from Bell Labs, Columbia University, the RAND Corporation, and IBM.[45] The first episode of this series, "Logic by Machine," begins with a crude animation of dinosaurs, followed by a loincloth-covered caveman, who, frustrated at not being able to eat a nut directly from a tree, smashes it with a rock. A voice-over informs the viewer:

> Man has described himself in many ways. But if we concentrate on man in relation to his control over his environment, no description is more apt than the description of man as a tool making animal. And in the short history of man, the majority of tools he has made can be thought of as an extension of his muscle power: that is: the ability to perform work faster and in greater quantity. However, in the mid-1940s of the 20th century, a different kind of tool was invented: a tool for extending certain of the powers of man's mind. This tool is the electronic computer.[46]

Framing man as an entity who has control over his environment and the computer as "a tool for extending certain of the powers of man's mind" places *The Computer and the Mind of Man* firmly in what Ronald R. Kline has called the "cybernetics moment," when many scientists across different disciplines began to believe that "information theory could bridge the physical biological, and social sciences," leading "scientists, engineers, journalists, and other writers in the United States to adopt these concepts and metaphors to an extent that is still evident today."[47] Though not mentioned directly in the first episode, cybernetics, the study of self-governing systems, was heavily thematized in later episodes in order to describe how computers function. In episode 3, "The Universal Machine," for example, the narrator declares: "We, the humans, can instruct the computer to imitate the behavior of any other specific machine or process, providing we can adequately describe what we want imitated, and can instruct the machine in its own terms how to carry out the imitation. [. . .] Classically, the control element is the human being himself [*sic*]. But in the digital computer, control is achieved electronically, in strict conformity with instructions provided the machine by the human programmer."[48]

To create his musical score for *The Computer and the Mind of Man*, Subotnick turned himself into a "control element" that was simultaneously human and computational, developing a novel technique that blurred the lines between human compositional activity and programmed performance.

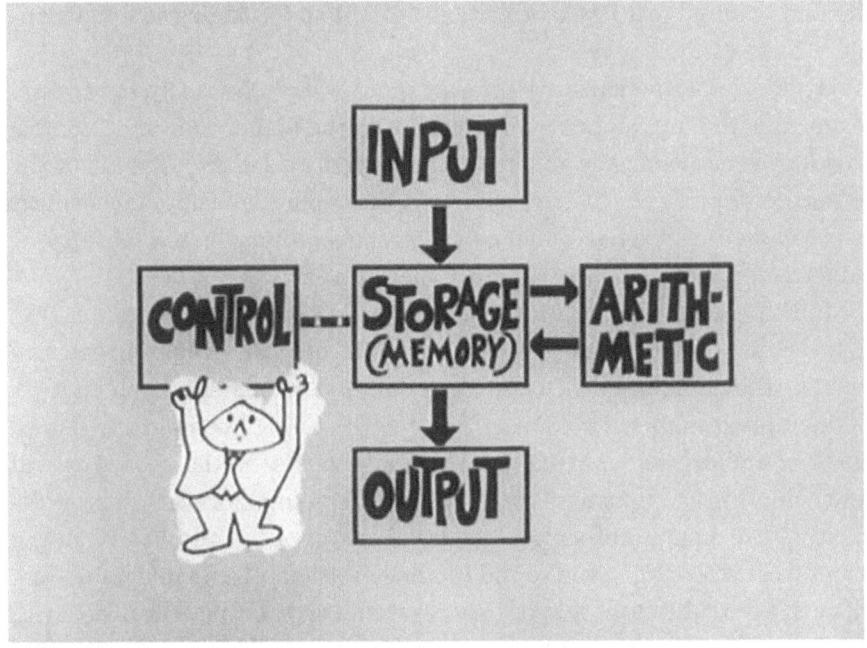

Figure 1. Still image from The Computer and the Mind of Man, episode 3, "The Universal Machine." source: KQED-TV/National Educational Television and Radio Center, 1962.

In his garage, he constructed a looping circuit of metal objects including streetcar pulleys, old piano frames, and salvaged scrap metal, placed a microphone in the center of the circuit, and placed a tape machine's transport controls at a single point in the room. The arrangement of the instruments in the circuit became a simple "program" that would be executed by Subotnick pressing the record button, running through the percussion circuit while hitting the instruments with drumsticks, and ending at the tape machine, where he would end the program by stopping the tape.[49]

Although one of the basic premises of *The Computer and the Mind of Man* was that a single human being could design a computational system that could in some sense control itself, the narrative reassured viewers that such control would be ultimately governed by its human programming. A simple diagram (see figure 1) illustrates this concept by inserting a cartoon human holding up a sign labeled "control" into a block diagram of a computer's constitutive parts.

At this point, the narrative informs the viewer: "Classically, the control element is the human being himself. But in the digital computer, control is achieved electronically, in strict conformity provided the machine by the human programmer. The great power of the computer resides in its ability to carry out whatever arithmetic or logical operations the mind of man is capable of instructing the machine to perform."[50]

In this understanding of computing, not only could a computer "extend certain of the powers of man's mind," it could also serve as an external model of the human mind itself, created by and ultimately beholden to its human programmer. The notion that a single human being could design a self-contained, self-controlling computational system that would extend their mind into a dynamic system echoed Subotnick's earlier "heroic vision" of the creation of a closed externalization of his mind through the careful control of light and sound in space. Yet Subotnick's understanding of such externalizations and enclosures would soon be again transformed, leading to his development of a concept of music specific to the new emerging media in the radically expanding technoscientific world of the 1960s.

UNDERSTANDING MEDIA

Subotnick's understanding of the power of music to extend the mind was soon transformed again through a mid-1963 encounter with a duplicated copy of Marshall McLuhan's 1960 *Report on Project in Understanding Media*, prepared for the National Association of Educational Broadcasters, which Subotnick has often referred to as the "holy grail."[51] By the time he encountered this document, Subotnick, along with the composer Ramon Sender, had established the SFTMC, which had settled into its most permanent home at 321 Divisadero Street in the Lower Haight; sometime in the summer of 1963, the poet, raconteur, and hereditary cheesemonger Gerd Stern had traveled to the SFTMC seeking out technical help for an upcoming show at the San Francisco Museum of Art, and he gave Subotnick a mimeographed copy of the report. Stern himself had obtained the copy from the poet M. C. Richards, his neighbor at the Gate Hill Cooperative in

Stony Point, New York; Stern speculated that Richards had been given a copy by John Cage.[52] With this provenance, Subotnick remembers being in the presence of an important, rarefied document, and recalls sitting on the floor overnight with Stern and others at the SFTMC, dividing up the manuscript into parts so everyone could read it simultaneously in one sitting.[53]

McLuhan's report boldly claimed that "any medium whatever is an extension, a projection in space or in time, of our various senses" and emphasized that it was concerned not with the contents of media, but rather with their effects.[54] For McLuhan, electronic media promised an end to the "aggressive, brawling sequence and cycle" of specialized media in the West—script, print, book, radio—continuously usurping and overlaying each other; instead of "one-thing-at-a-timeness," McLuhan saw electronic media as promising "all-at-once-ness."[55] This was because McLuhan understood electronic media to deal with pure information, creating a speculative "global village" in a world now connected at the speed of light.[56] Subotnick remembers McLuhan's report as finally articulating the ideas he had been developing for the previous several years through his work with technology, reinforcing and legitimating his notion that individual composers could "extend" themselves in sensory space and time through new electronic technologies.

As Fred Turner has argued, the 1964 book that emerged from McLuhan's 1960 report, *Understanding Media: The Extensions of Man*, marked a popular articulation of a decades-long trajectory of ideological work on the part of American social scientists, politicians, and artists that led to the creation of what he has called "democratic surrounds": "multi-image, multi-sound-source media environments" that modeled "a way of thinking about organizing society" around ideals of individuality, diversity, and democratic cooperation.[57] The origins of this ideological project, Turner has shown, lay in American suspicions of the connections between mass media and the rise of fascism in the 1930s and 1940s, especially in relation to propaganda distributed by the Nazi regime. To counter the effects of mass media, organizations such as the Committee for National Morale explicitly sought to promote individuality using models of an individuated personality from both behavioral psychology and psychoanalysis. Committee members such as the psychologist Gordon Allport defined *personality* in 1937 as "a dynamic organization within the individual of those psychophysical systems that determine his unique adjustments to his environment";[58] by framing personality as a cybernetic, self-regulating and adaptive system, Turner shows, Allport "allowed the individual to emerge as a creative being," fostering a sense of liberality that both freed the individual from the deterministic, psychoanalytic inheritance of their childhood development and positioned them as being open to change through "shifts in [their] social and symbolic surroundings."[59]

Such a liberal conception of the dynamic, individual personality must have been attractive to Subotnick, who has long discussed his desire to unburden himself of the enculturation he underwent as a young person into the world of classical music as a clarinet virtuoso.[60] And indeed, in the period immediately surrounding Subotnick's encounter with McLuhan's manuscript, he collaborated on many projects with fellow members of the SFTMC and the Dancers' Workshop, run by choreographer Anna Halprin, that could be described as "democratic surrounds." Chief among these was an installation called the *Environment and Sound Mobile*, which premiered on May 26, 1963, on the campus of Mills College, where Subotnick was employed, during a festival that nominally celebrated the seventieth birthday of the composer Darius Milhaud.[61]

THE *ENVIRONMENT AND SOUND MOBILE* AND THE SPECTER OF THE HAPPENING

The *Environment and Sound Mobile* was a collaboration between Subotnick, Ramon Sender, and the sculptor Robert Dhaemers, who described his work as "organic modernism."[62] Evoking the kinetic sculpture of Alexander Calder and the sonic sculpture of Jean Tinguely, Dhaemers's sculpture for this piece consisted of a large metal facade held up by wooden supports, holding multiple panels, industrial parts, gears, escapement mechanisms, levers, and other items crudely welded together. Some parts, such as a clock face and a small fan, moved, but it is impossible to tell from the archival documentation if these sculptural parts themselves produced sound. However, the work's documentation does show that the sculpture's facade hid several tape machines and amplifiers that fed loudspeakers interspersed throughout the Mills College campus, allowing the object to produce sound for several hours without human intervention.⁶³ Local television station KQED-TV's coverage of the festival described the *Mobile* as "a kind of outdoor concert involving electronic music"; sounds included on their news segment cut between a laughing audience, a series of upward harmonics on a flute, reverberant pitched bells played at random, piano glissandi, and other abstract sounds.⁶⁴

The *Mobile* was only one element of this environment, however. The performance also involved the visual and haptic dimension of dancers who were stationed throughout Mills's campus in balconies, windows, and fields, each wearing nontraditional outfits and executing movements similar to the everyday choreography developed by Anna Halprin for the Dancers' Workshop. An article in *Time* magazine about the event called it a "happening," emphasizing the apparent novelty of this work during a festival intended to honor Darius Milhaud through traditional performances of musical works and connecting it to the kinds of multimedia events popularized by Allan Kaprow (and, as Turner has argued, by other students of John Cage in the 1960s interested in indeterminacy).⁶⁵

However, the program note for the *Mobile*, most likely written by Subotnick, goes to great lengths to specify that the work was not a Happening, in which an individual attendee could determine their own experience and make their own meaning. Instead, it insists that the *Mobile* was an experimental attempt to bring together multiple media for the simultaneous performance of distinct, discrete works by individual authors. Closely following the style and argument of McLuhan's 1960 *Report*, this note avoids discussing the content of the work and instead emphasizes the effects of its individual media:

> The magnetic tape has become a new medium for composers. It has added new sound materials to the language of music and a revolutionary method of presentation. The fact that a musical composition can be created and presented without the aid of musicians changes dramatically the concert hall atmosphere. There is a kind of "pure" electronic music which does not specify the mode of presentation beyond the kind of tape recorder and the number of speakers. But there is also a growing attempt to evolve a new concept of public presentation for the new medium. The *Environmental and Sound Mobile* is a collaborative attempt to combine tape music with the visual arts. [...] The visual environment is designed to add another dimension to the experience of the moving audience.⁶⁶

Though the *Mobile* was an inherently collaborative project—involving Subotnick, Sender, Dhaemers, a choreographer, and many unnamed dancers—this note frames it as a musical composition onto which other works had been added in different sensory modalities, and therefore different media. Indeed, it describes an entirely separate "visual environment," indicating that the use of the term *environment* differed significantly from its use by artists such as Kaprow, whose contemporaneous *Environments* series of installations sought to create complete worlds. Instead, this note suggests that each sense modality, and each corresponding medium, could have its own mono-sensory "environment." The note also theorizes magnetic tape as a "new medium for composers," affording a new kind of cultural production in which a single "musical composition" could be created and presented without "the aid of musicians"—without any other human agency beyond the composer's. Thus the medium of magnetic tape was fantasized as an extension of the composer's creative will, even in situations where it was combined with other technologies, other instruments, and other media.

This program note hints at the discomfort Subotnick felt with framing multimedia performance as a democratic "happening," preferring to interpret McLuhan's thesis as laying the groundwork for a new medium-specific "'pure' electronic music."[67] In a 2013 interview with Frank Oteri, Subotnick candidly recalled a feeling of dissatisfaction with all of his early collaborative works of this period, beginning with *Sound Blocks* and continuing to the period of the *Mobile*: "No, I don't know what the hell I'm doing. I don't know what this is that I made. I've got to re-think this whole thing."[68] Blau was similarly dismissive of *Sound Blocks*: "One of the hazards of collage is its prodigality—the difficulty of resolve in the art of waste. [. . .] It isn't quite a Happening, but doesn't know why it shouldn't be."[69] Subotnick's peers at the SFTMC also dismissed Happenings, or at the very least sought to differentiate their work; although the SFTMC's visual director Anthony Martin provided visual art for most SFTMC concert performances, and indeed would go on to create installation work that he called "environmental" later in the 1960s, he never read McLuhan and considered his work to "complement" other work presented at the same time.[70]

By 1964 Subotnick explicitly differentiated his work from Happenings by envisioning two ways forward after reading McLuhan. One was the "'pure electronic music" formulated in the program note of the *Mobile*: a new electronic music that could singularly extend the composer's mind. The other was a "live performance" environment in which collaborators could create "parallel pieces" that could form a "counterpart" to his work.[71] In both cases, Subotnick's electronic music would be "purified" from other collaborators; the fact that it was electronic, in addition, meant that it also could be purified from other media. Indeed, while McLuhan's 1964 book *Understanding Media* asserts that "electric light is pure information," thus excepting it from the dictum that "the content of any medium is always another medium," his earlier 1960 *Report*—the document read by Subotnick—also asserted that instrumental music, along with "non-objective art," was similarly contentless. "Our illusion of 'content' derives from one medium being 'within' or simultaneous with another," it reads. "For this reason instrumental music has no 'content' and non-objective art likewise is an abstract manipulation of the modalities of sight."[72]

McLuhan's bold claim about the abstract neutrality of instrumental music, which was eventually excised from *Understanding Media*, recalls the claims made by Milton Babbitt about the abstract neutrality of electronic instruments such as the RCA Synthesizer, which he had demonstrated to Subotnick in 1959. Instead of composing musical "content" determined by the social and historical "mediations" of the autonomous, liberal human composer's imagination—musical scores, orchestras, concert halls, and so forth—new electronic instruments promised to directly manipulate the human perceptual apparatus of sound. Although Subotnick had seen the RCA Synthesizer in action, it did not satisfy his desire for an instrument that could achieve such direct extension of the composer's mind. Not only was it extremely tedious to program, it also forced its users to specify their desired sound events using historically and socially determined parameters, including twelve-tone equal temperament, a steady tempo and standard rhythmic subdivisions, and quantized settings for other parameters such as timbre.[73] The user would also have to gain an intricate and intimate knowledge of the instrument's inner workings, since each coded binary instruction would be sent to a module that would have to be manually adjusted for each desired sound. Subotnick wanted something simpler, an instrument that would allow him to input his imaginative creativity simply and directly into the electronic medium and immediately output sound.

TOWARD THE COMPOSER'S BLACK BOX

Subotnick remembers that beginning in 1964, he placed classified advertisements in the *San Francisco Chronicle* seeking out an engineer who could make his new medium-specific instrument for electronic music, even though he had no idea what it would be. To Subotnick, this composer's black box would radically break with the ways he had previously attempted to create electronic music, including both the editing together of various pieces of magnetic tape to create *musique concrète* and using extant "synthesizers" (including both electronic keyboard instruments and the RCA Synthesizer). As Subotnick remembered: "We knew that it was the beginning of an era. We knew that those synthesizers were not the most powerful instruments that could be created. We also knew that the era was evolving beyond pure studio work like cutting tape."[74]

Although Subotnick had been working at the SFTMC since 1962, cutting and splicing tape to create electronic music was not satisfactory to his ambitions. At the center, Subotnick had access to tape machines, oscillators, and various other ad hoc instruments that are discussed in more detail in chapter 2, but none of them was a purpose-built, medium-specific instrument for the kind of electronic music he was imagining could be possible through new technology. Like many others at the SFTMC, Subotnick was also interested in moving away from creating electronic music through the laborious processes of recording, cutting, splicing, and editing tape asynchronously; he remembers thinking that "what I was working with was not the way to go—cutting tape wasn't it. So I started to look for something that would be more meaningful, to be able to be in the studio painting with sound."[75]

Along with Ramon Sender, Subotnick began to imagine a new electronic musical instrument that would metaphorically allow him to "paint" sound. This initial speculative instrument was a device that would generate electronic waveforms using a rotating disk with a unique pattern of holes punched in it, a light source, and photoresistors. As Subotnick remembered in 2017, the process for creating this instrument would consist of the following steps: "Create a pattern of holes on a flat round disc; spin the disk with a variable speed motor; pass light through the rotating disc; convert the resulting light pattern to sound by placing a photocell to receive the light pattern passing through the disc. A pattern could be made for each sound; the size of the pattern would represent amplitude; the shape would result in timbre and the speed of rotation would be some kind of frequency change."[76]

Subotnick's proposal for a purpose-built electronic musical instrument was hardly unique, however, and reveals how the ideological promise of electronic sound held power over him even without a working technical knowledge of electricity, electronics, or instrument design. Although Subotnick cited the work of the filmmaker Norman McLaren as the inspiration for this design, it is more indebted to the older, more general technology of the perforated disc. McClaren's "drawn-on-sound" technique, which he began developing in the 1950s, involved drawing a soundtrack directly on 35 mm film stock with black ink, essentially producing crude pulse waves with little precision in specifying frequency, amplitude, or timbre.[77] Perforated discs like the one described by Subotnick were a much older technology, notably used in the acoustic siren, one of the first devices that allowed for the precise measurement of the fundamental frequency of sound waves; they also formed the basis for the Rhythmicon, an instrument built by Leon Theremin for Henry Cowell in 1930 that used a perforated disc and photocells to create unique rhythmic patterns that articulated the rational relationships of the harmonic series in both pitch and rhythm.[78]

Subotnick's lack of technical knowledge seems to have enhanced his imagination for what an electronic musical instrument could do. By envisioning such an instrument as a composer's black box, he did not need to know how it worked, only that it took the individual's musical creativity as input; freed it from historical, stylistic, and formal restraints; and produced electronic sound as the output of that imagination. In a 2017 interview, Subotnick summed up his position on the latent potential of human musical creativity: "[E]verybody is potentially musical and has the capacity to appreciate music. It's like almost anyone could learn to play chess, but to be able to strategize and be really good at it requires a kind of concentration that very few of us have. Composers often have a kind of advanced skill at hearing things and putting things together. So maybe not everyone can be a 'composer,' but it doesn't mean that everyone doesn't have the ability to compose."[79]

With a sentiment that recalls Babbitt's liberal attitude toward the RCA Synthesizer, Subotnick conceptualized musical creativity and appreciation as basic cognitive functions akin to walking or playing chess, implying that musicality is not something one is born with, or something learned socially, but rather something that could be developed from basic cognitive capacities. Comparing the creation of music to playing chess, Subotnick invoked a classic example of a cognitive skill that became offloaded to a computational intelligence, rendering the social, historical, and political markers of that body's identity invisible. Through the use of computerized interfaces that apparently mirrored and externalized the internal cognitive processes of musicality, Subotnick imagined a simplification of "composing" into gestural input and sonic output, allowing any person to be musical without training or skill. Such a composer's black box would enable anybody to paint music on a music easel.

But as Subotnick would quickly discover, black boxing human musicality would not only make possible the metaphor of "music as a studio art" that he hoped would fulfill his vision of a pure electronic music—that is to say, an electronic music unique to the individual human who created it, using their own unique black-boxed instrument, which extended the mind of the composer directly to the listener—but also something more. Shortly after Subotnick imagined his first composer's black box, he encountered the engineer Donald Buchla, who stopped by the SFTMC to use their three-track tape machine. As explored in the following chapter, the box that Buchla eventually made and delivered to the SFTMC was much more than what Subotnick imagined it would be. It was a complex, nonlinear, dynamical system that operationalized a computational model of the human mind that differed from Subotnick's in significant ways. For Buchla, the human mind was not the core of a bounded, static, liberal subject that could extend itself into the world, ingeniously creating and controlling a black box to do its bidding. Instead, for Buchla, the human mind itself was also a black box, a complex cybernetic system of various physiological and cognitive processes, which could be explored, and expanded, through psychedelic technologies—technologies that promised the manifestation of the musical and creative mind, opening the doors of perception to subjectivities unknown.

2 Opening Buchla's Box

MANIFESTING THE MUSICAL MIND

In an early 1964 issue of the *San Francisco Chronicle*, nestled between classified advertisements placed by IBM, Motorola, Fairchild, Honeywell, and Westinghouse for electrical and systems engineers, a small ad sought an engineer to build what its author, the composer Morton Subotnick, called an electronic "music easel." (Or so he remembers.)[1] With such a device, Subotnick fantasized, any human being, with or without musical training, could "paint" music electronically, transforming it into "studio art" in which the human artist had complete control over their work from ideation to material realization. Though Subotnick did not know how such a device would work, or what exactly it would produce, he did know what he wanted it to be: a direct electronic conduit between the innermost creative subjectivity of a human being and sound, bypassing the intervening steps of writing a score and finding musicians to play it. To use an engineering term of his day, such a music easel would be a black box, collapsing the tedious complexities of the creation of electronic music into a simple scheme of input and output. In this metaphorical fantasy, the human user would make a musical gesture akin to painting, and without needing to understand how the instrument worked, their music easel would output what Subotnick imagined as their "personalized music."[2]

The fantasy of the gestural painting of sound was not new. What was new, however, was that Subotnick's speculative instrument was, as he came to call it, a composer's black box, wherein such gestural painting would take place through the abstracted domain of electronic information, rather than ink or paint.[3] Indeed, in most other cases where musicians imagined instruments that would allow them to draw sound—including Percy Grainger's "Free Music Machine" (1892–1948), Yevgeny Murzin's ANS Synthesizer (1957), and Daphne Oram's "Oramics" machine (1959), among many others—the human operator would need to understand how the machine worked in order to use it. Contrastingly, in Subotnick's imagination, not only was such technical knowledge unnecessary, but it would also serve as a distraction from the device's true innovation of allowing a human to extend themselves directly and instantaneously through electronic sound; his speculative easel required no canvas, brushes, or palette. Such a direct, electronic extension of the self was one of Subotnick's main goals both as a composer and as a philosopher of music, emerging from his encounter with the nascent media theory of Marshall McLuhan in the early 1960s. As McLuhan promised, electronic media were "simultaneous," extending the human sensorium to create a "unified field of experience."[4] That such a simultaneous and unified field of experience could be directly transmitted from a single composer's mind to the auditory apparatus of a listener was, for Subotnick, the ultimate goal of his speculative instrument.

Subotnick's desire for such a speculative instrument was both ideological and practical. In theory, information moved at "approximately the speed of light" in McLuhan's brave new "electronic age," but in practice, the electronic media with which Subotnick had been working up until 1964 moved at the speed of mud.[5] To record one second of electronic audio required between seven and thirty inches of magnetic audiotape, which meant that making a concert-length "tape piece" required months of laborious splicing. Working with oscillators, electronic switches, and other laboratory test devices to produce electronic sounds to record onto tape was no easier. Designed for use in radio, telephony, and television, such devices were difficult to operate, especially for a musician with no knowledge of electronics or even electricity. And for Subotnick, making electronic music in 1964 was even more difficult because the studio in which he was working was barely a studio at all.

The SFTMC, which Subotnick cofounded in 1962 with the composer Ramon Sender and later ran with both Sender and the composer Pauline Oliveros, was an ambitiously named organization that telegraphed its association with *tape music*, a novel term that valorized the technicity of magnetic audiotape over the European stylistic schools of *elektronische Musik* or *musique concrète*.⁶ But despite its name, the SFTMC consistently struggled to build an operational studio, presenting far more music than it produced. Without any technical expertise or institutional funding, the center's studio floundered for its first two years of existence; even as late as the summer of 1964, one member-composer recalled that "there wasn't much of a studio [...] There was a monaural tape recorder and a Hewlett Packard sine wave oscillator. Oh, and I think there might have been an equalizer."⁷ Various other members recalled a motley assortment of equipment sourced from World War II bombers, railroad salvage, and the University of California's "Department of Physics and Cosmic Rays."⁸ William Maginnis, the center's second technician hired in the summer of 1964, fondly remembered the studio as "the great grand kludge!"⁹ Despite Subotnick's embrace of McLuhan's motto "the medium is the message," the medium was still tedium.

Subotnick's music easel promised to take that tedium and turn it into something that its human user didn't have to think about. As a black box, it would ostensibly allow for direct, immediate access to a purified medium of electronic sound. Yet in seeking out an engineer in the San Francisco Bay Area in 1964 to create such a black box—in a time and place in which black boxes were being developed in the service of cybernetic, self-governing systems—Subotnick got much more than he bargained for. Shortly after placing his classified ad, a young Berkeley-trained engineer named Donald Buchla showed up at the SFTMC to use their three-track Ampex tape machine and was readily conscripted to build Subotnick's black box. Buchla was no stranger to such boxes: trained during the height of the Cold War, he had built many similar devices at the Lawrence Radiation Laboratory, colloquially known as the Rad Lab, across the Bay in Berkeley. Buchla's earliest black boxes took as input the biological functions of animals and produced as output electronic signals, creating data that would be used to develop self-governing environments to sustain "unrestrained primates on extended space flights."¹⁰ Steeped in the cybernetic systems engineering of the Cold War space race, Buchla understood a human being not as a bounded, individual subject that could be extended through electronic technologies, but rather as a complex conglomeration of systematic processes that could itself be radically transformed through those same technologies.

Although Buchla shared Subotnick's interest in the biological production of electronic signals, his approach was informed by a deep commitment to the psychedelic exploration of human consciousness, achieved with technologies both electronic and chemical. As an early and devoted user of such drugs as lysergic acid diethylamide (LSD), supplied to him by his associate Augustus Owsley "Bear" Stanley III beginning in 1965, Buchla was interested in how human consciousness and sensation—particularly auditory sensation—might be explored and expanded through tinkering with the chemical and sensory inputs into the human brain, altering its internal configuration. For Buchla, building a composer's black box meant more than collapsing the complexities of the creation of electronic music into input and output; it meant constructing an architectural system that was complex, nonlinear, and cybernetic, just like his understanding of the human brain and the human being as a whole. Rather than imagining a composer's black box as extending an individual composer's subjectivity through electronic sound, Buchla imagined the composer to be a black box themselves.

When Buchla eventually delivered his box to the SFTMC sometime in late 1965, people simply called it the Buchla Box.[11] Consisting of two wooden cabinets packed with separate modules, with two custom-built controllers that responded to human touch, it presented its human user with a computational architecture for creating "patches" of electronic signal that would travel between skin, metal plates, cables, and circuitry, allowing for the emergence of automatic, cybernetic self-control. Although this simplified the input and output behaviors of the system for its user, black boxing the complexities of parametric control by making them unnecessary to understand, it also distributed control among each module in the system, complicating the fantasy of the direct and total control of electronic sound by the system's human user. Buchla and his associates used his boxes for a variety of experimental and psychedelic purposes throughout 1966, but it was not until 1967, upon the release of Subotnick's album

Silver Apples of the Moon, that the Buchla Box gained widespread popularity as what the album's jacket described as an "electronic music synthesizer," and because of the popularity of a morphologically similar system of interconnected modules designed by Robert Moog in Ithaca, New York, the Buchla Box later became glossed as "the earliest analog voltage-controlled synthesizer."[12]

As this chapter shows, however, the Buchla Box was designed neither as a sound synthesizer nor as a composer's black box that would allow its human user to directly extend their compositional ideas through a purified medium of electronic sound. Instead, the Buchla Box was designed as an externalized model of the human black box: a complex, nonlinear system of interrelated functions whose creative potential only emerged through the act of real-time performance. In other words, rather than extending a knowable, stable, and bounded human subject, the Buchla Box positioned its human user as only one agent in an emergent cybernetic system. This posed a challenge for the social identity of the human composer as the ultimate source of musical material and to the understanding of musical instruments—even instruments such as the Buchla Box—as "tools" at the composer's disposal for expressing their internal creative ideas.[13] Indeed, as this chapter explores, the earliest users of Buchla's Boxes were explicitly interested in what I provisionally term "ex-compositional" uses of this new technology, creating musical performances that expanded human consciousness itself through the distribution of creative agency and control between the Buchla Box and its human user.

To map the emergence of the Buchla Box at the SFTMC, this chapter begins by discussing Buchla's position as an electrical and systems engineer in the Cold War cybernetics moment, designing black boxes for biometric telemetry and auditory prostheses for the blind. It then closely follows Buchla through the SFTMC, describing the state of the center's technological apparatuses before and after his arrival and showing how the various modules designed for the Box emerged from the center's idiosyncratic technological contingencies. It then attends to the earliest uses of Buchla's Boxes, including improvised performances by the center's technician William Maginnis, as well as their use among the Bay Area's nascent counterculture. Ultimately, it argues that the Buchla Box served as a technology of consciousness that challenged inherited notions of musicality, producing emergent, often ephemeral, cybernetic performances that largely remained illegible in the institutional social world of composition, and which were ultimately overshadowed by its retroactive incorporation into the emerging discourse of the synthesizer in the 1970s and beyond.

BUCHLA'S EARLY SYSTEMS: BIOMETRICS AND CYBERNETICS

Buchla's earliest boxes, built in the throes of the cybernetics moment in the early 1960s, turned animals, and humans, into informational beings. Educated in physics between 1955 and 1961 at the University of California, Buchla was among a generation of American engineers trained to meet the demands of the atomic age, in both the space race and the more earthly race for nuclear arms. In Berkeley, Buchla worked at the Lawrence Radiation Laboratory, a national center for research on radiation and particle physics, where he not only studied as an undergraduate but also was employed as a freelancer after graduating. But although Buchla earned a living contributing to what the social theorist Theodore Roszak came to call the "technocracy"—a society increasingly determined by technical expertise, which perniciously developed a subliminal totalitarianism in everyday life—he also was a member of what Roszak termed the "counterculture," an antiestablishment, liberal culture that was "radically disaffiliated from the mainstream assumptions of our society."[14]

With his interest in experimental electronic music and penchant for psychedelic drugs, Buchla continued a long Californian tradition of the exploration of sensation and consciousness, well before the neologism "hippie" was coined in 1965.[15] Connecting Buchla's countercultural interest in exploring consciousness and his work for Big Science was the technoscientific interdiscipline of cybernetics, which valorized self-regulation, emergent dynamics, and complex systematicity. Although his work made it possible to conceptualize humans as systems of flowing information that could be controlled by top-down state apparatuses, those systems also created possibilities for bottom-up, emergent, and radically individualized control. Indeed, as Fred Turner has argued, cybernetics formed a bridge between Roszak's categories of the technocracy and the counterculture, connecting them through their mutual interest in the flow of information: "For much of the broader counterculture, cybernetics and systems theory offered an ideological alternative [to technological bureaucracy]. Like Norbert Wiener two decades earlier, many in the counterculture saw in cybernetics a vision of a world built not around vertical hierarchies and top-down flows of power, but around looping circuits of energy and information. These circuits presented the possibility of a stable social order based not on the psychologically distressing chains of command that characterized military and corporate life, but on the ebb and flow of communication."[16]

Buchla's projects at the Rad Lab very concretely created such "looping circuits of energy and information." Throughout the early 1960s, Buchla worked on several bioinformatics projects for the National Aeronautics and Space Administration (NASA), designing and constructing "surgically implanted, biologically powered transmitters" that measured and transmitted electronic signals from the respiratory, circulatory, integumentary, musculoskeletal, and nervous systems of primates, rabbits, and other animals.[17] Described on a later curriculum vitae as "monitoring physiological parameters in unrestrained primates on extended space flights," these experiments involved the vivisection of live animals, who would have electronic devices implanted into various parts of their bodies and then be cruelly exposed to extreme environmental conditions until they perished.[18] The information produced by these experiments was used explicitly toward the goal of maintaining life in the harsh environmental conditions of outer space, both through cybernetically controlling the environment in which the animals found themselves and, more uniquely, through cybernetically controlling the biological functions of the animals themselves. Such "cybernetic organisms," or "cyborgs," as Manfred Clynes and Nathan Kline termed them in 1960, were imagined as "self-regulating man-machine systems" that would "leave man free to explore, to create, to think, and to feel."[19]

Buchla's early bioinformatics projects built on a long trajectory of what Peter Galison describes as "the elision of the human and non-human" that underpinned cybernetic thought throughout the twentieth century.[20] Such a "blurring of the machine-human boundary" was essential to the articulation of cybernetics by Norbert Weiner, who, for example, wrote in 1948 that "we are beginning to see that such important elements as the neurons, the atoms of the nervous complex of our body, do their work under much the same conditions as vacuum tubes."[21] One of the major shifts with the cybernetic organisms imagined by Clynes and Kline beginning in 1960 was that biological systems could not only be electronically measured, but could also themselves be radically transformed through cybernetic engineering. The potential of such systems, as Stefan Helmreich notes, was both electrophysiological and ideological, pointing "both to putatively 'natural,' possibly teleological, forces considered to be latent within organisms as well as to those possibilities that might be socially realized as people select how to direct such forces in the near and not so near term."[22] The electrical potentials of neurons in the brain were easily metaphorized as social potentials for new kinds of cultural, political, and creative activities in the 1960s.

One major social potential that Buchla saw in the generation of bioinformatic signals, as well as the blurred boundary between humans and machines, was to expand cognitive connections between human sensory apparatuses and human consciousness. In 1961, Buchla began to develop a device called the ORB (Optical Ranging for the Blind), which, according to a later curriculum vitae, "sparked a continuing and relevant interest in the refinement and utilization of communications channels between man and electronic system."[23] The ORB was a handheld device that produced an audible signal whose parameters would be mapped to an object's given proximity. The ORB's cutting-edge technologies included rotating sets of mirrors, lasers, first-generation light-emitting diodes (LEDs), and miniaturized amplifiers and loudspeakers that operated with both sound and ultrasound.[24] Buchla explicitly envisioned this device as expanding the cognitive potentials of the blind: "I found that congenitally blind people had enormous ranges for learning, enormous centers for basically doing pitch discrimination and acoustic analysis of their environment."[25] The ORB became a prosthetic extension of an imagined sensory system with a narrowly defined "deficiency" (blindness), and modified that system through the cybernetic intervention of sonic feedback.[26]

Electronic sound was a source of potential—electronic, technological, and social—for Buchla. As he experimented with sound as a source of real-time feedback to externalize and expand human sensory processes, he also experimented with sound recorded onto audiotape in a compositional practice he located in the tradition of *musique concrète*, creating tape works at home with his one-track Wollensak tape recorder.[27] Buchla's interest in the burgeoning discourse of tape music drew him across the San Francisco Bay to the SFTMC, which since 1962 had been presenting high-profile concerts of experimental music that utilized novel electronics. More concretely, Buchla was interested in using the SFTMC's unique three-track Ampex tape machine, which had been featured in publicity photographs published in the *San Francisco Examiner* in March 1964.[28] In his memory, completely unaware of Subotnick's classified ad seeking an engineer to build a composer's black box, he simply attended a concert at the SFTMC's space at 321 Divisadero Street sometime between late 1963 and early 1964, and somehow convinced Subotnick to let him use their Ampex machine.[29] But like Subotnick, Buchla also got more than he bargained for in this encounter. At the SFTMC, he was quickly enlisted to fix the studio's kludged-together equipment, and in so doing, he began to develop the kernel of the idea that became the Buchla Box: a single modular system that would standardize and systematize the functions of an electronic music studio, creating the conditions for the emergence of a new kind of cybernetic musical creativity.

"THE GREAT GRAND KLUDGE!"

At the SFTMC, Buchla met a community of composers for whom the technoscientific discourses of cybernetics and information theory had generated a radically different set of ideas than those that motivated his own work. As discussed in chapter 1, Morton Subotnick's desire for a composer's black box that would serve as an easel on which composers could more easily create electronic music was essentially a desire for a new instrument that simplified and black boxed the process of creating music, immediately extending the composer's musical creativity into electronic sound.[30] Yet at the time of Buchla's visit, none of the projects initiated by members of the SFTMC had produced a satisfactory black-boxed musical instrument. The membership of the SFTMC included people trained as composers, amateur musicians, and amateur electrical engineers from around the Bay Area, but none had any extensive technical expertise. Even the studio's flagship three-track Ampex tape recorder was barely functional; hand built by a high school student who had briefly interned at Ampex, it ran slightly slower than other machines, and any attempt to repair it would cause its tape heads to fall into dozens of tiny pieces, since they were never properly assembled in the first place.[31] And when the center finally acquired a reliable set of tape machines—a slightly damaged Ampex PD-10 tape duplication system that had fallen off a shipping pallet, discussed more in chapter 3—its members still faced the ongoing challenge of the amount of labor it took to cut and splice tape.

Ironically, tape proved to be the bane of the SFTMC. An early attempt in 1963 to simplify the sequential or even simultaneous playback of multiple segments of audiotape involved retrofitting an organ keyboard with electromechanical relays mounted to a large plywood board underneath its keys, which would in turn switch on separate tape transports. This project was given to the center's first technician, Michael Callahan, an amateur engineer who had begun working with Sender and Oliveros as a high school student in 1961 during their *SONICS* concert series at the San Francisco Conservatory of Music (SFCM), and who continued to work at the center until the summer of 1964, when he joined Gerd Stern and Steve Durkee to form the multimedia art collective the Company of Us (USCO). Although Callahan did produce a prototype, it was both figuratively and literally clunky; the large railroad surplus relays produced loud clunks when activated, producing noisy distortion when recorded to tape.[32]

The center's second technician, William Maginnis, began work almost immediately after Callahan's departure in mid-1964; he wandered into the center to ask if he could use their ring modulator, and upon hearing him say the words "ring modulator," the Center's co-director Ramon Sender simply handed him the keys to the building and told him he was the new technician without even asking his name. In Maginnis's memory, Subotnick had traveled to visit the CPEMC in the summer of 1964 and returned to San Francisco to ask Maginnis to build a "sequencer" that would store and execute a preprogrammed sequence of events, such as the switching on and off of multiple tape machines, inspired by equipment he saw in New York. Maginnis's first idea for this project was to control a set of tape transports with a surplus telephone stepping relay system and a rotary telephone dial, such that a user could dial a number to turn on a corresponding tape machine. But as both the dial and relay were built for telephony, there was no real way to control the timing or other parameters of either device, and the project quickly stalled.[33]

Buchla was able to quickly and efficiently fix many of the SFTMC's kludges through his engineering expertise and access to professional fabrication equipment at the Rad Lab.[34] He replaced the clunky electromechanical relays of the center's first keyboard experiment by substituting resistive opto-isolators, which paired lightbulbs and cadmium-sulfide (CdS) cells, photoresistors developed in 1954 that were used in applications ranging from consumer electronics to micrometeorite detectors for NASA.[35] To help with a visual waveform generator project initiated by Sender, Buchla brought in a specialized high-voltage oscilloscope designed to produce high-intensity images that could be photographed. Though it apparently worked perfectly, Maginnis was wary of the X-ray radiation he realized was being emitted by the unshielded 40,000-volt cathode ray tube, and the project was quickly halted.[36]

Throughout 1964, Buchla brought his own prototypes to the center, such as the ORB, which he remembers showing to John Cage during a series of concerts at the SFTMC in March 1964, and which Ramon Sender remembers placing in the window display of the City Lights bookstore.[37] But while Buchla's fixes made these individual instruments operational, they did not solve the fundamental desire of many of the center's musicians for a more immediate way to work with electronic sound. With a thorough understanding of the various needs of many different types of musicians and an applied knowledge of the particularities of the SFTMC's kludged-together studio, Buchla most likely realized that what the SFTMC needed was not a set of unique, individual instruments for each musician's desires. Instead, it needed a system that would standardize and operationalize those desires into discrete electronic functions, separated into different modules that could be endlessly reconfigured to explore new relationships between the human user, the electronic instrumentation, and the common electronic signal that promised to connect them in new, emergent, and exploratory ways.

TOWARD THE BUCHLA BOX

Beginning at some point in 1964, Buchla began discussing with Subotnick and Sender the possibility of a centralized, modular system for controlling the functions of the SFTMC's studio. Subotnick remembers that during these initial meetings he explained to Buchla his concepts of music as a studio art that could be attained with a music easel. Sender's goals were much more modest: he was still only seeking a way to control the playback of prerecorded audiotape. Indeed, Sender himself had been modifying a Chamberlin Music Master, a novel instrument whose musical keyboard activated the playback of sixty-six individual loops of 3/8-inch magnetic audiotape, to play his own tapes. While Sender pushed Buchla to incorporate the Chamberlin and its musical keyboard into his system, neither Buchla nor Subotnick supported this idea, and in the memory of Anthony Martin, the SFTMC's visual director, the Buchla Box quickly became "Mort's project."[38]

To concretely operationalize his abstract desire for a music easel, Subotnick turned to a perhaps surprising model: Pierre Boulez's 1955 chamber cantata *Le Marteau sans Maître*.[39] Using this work as a model for the composer's black box was surprising not only because Subotnick did not use serial techniques in his own compositional work, but also because *Le Marteau* did not involve any sort of electronic or electroacoustic sound. Despite these incongruities, however, *Le Marteau* was widely understood to have been composed using incredibly complex, nearly opaque, serial operations, which for Subotnick perhaps modeled the compositional complexities he was interested in black boxing with his music easel. Indeed, as Stephen Heinemann has observed, Boulez utilized a unique serial technique of pitch-class set multiplication in this work that was not generally understood until the work was analyzed in the late 1970s. As Heinemann summarizes, in Boulez's technique, "the operation is arbitrary; the results are unpredictable; compositional and analytical choices are capricious; and important properties rise from arithmetical, as opposed to pitch-class, operations."[40]

During his meetings with Buchla throughout 1964, Subotnick remembers using the first page of *Le Marteau*'s score to specify the operations he would like Buchla's system to be able to execute: "I would imagine patching the Boulez and see something was missing and we would add another knob."[41] Using *Le Marteau* as a model suggests that Subotnick's vision for his composer's black box was not of a musical instrument that could be played in real time, but rather of a kind of musical computer that would allow a composer to input the values of musical parameters such as pitch, rhythm, note duration, dynamics, or even possibly timbre, and which would output a sound recording of an electronically realized version of a composition.

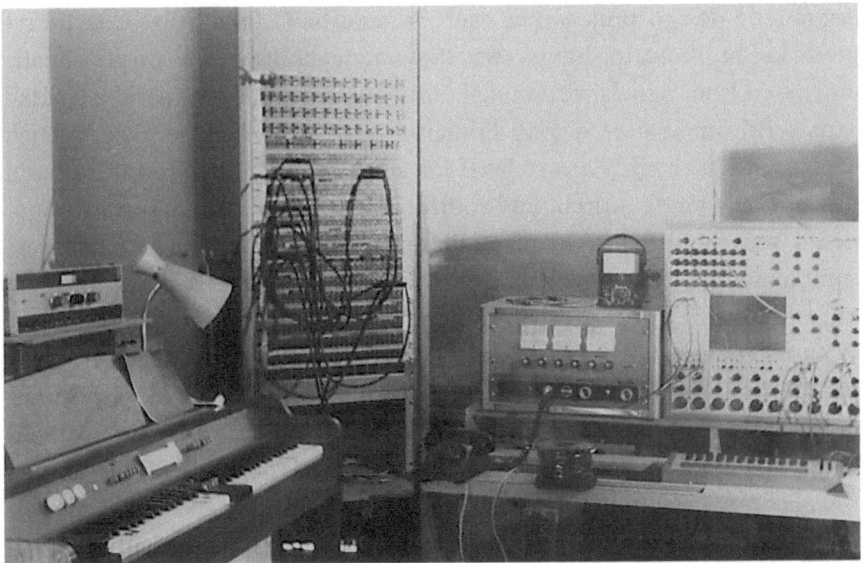

Figure 2. A partial view of the Buchla Modular Electronic Music System (right) at the SFTMC, 1965(?). Photograph by William Magennis.

In Subotnick's memory, the friction between his music-compositional desires and Buchla's expertise in electrical engineering drove Subotnick to research basic literature on both electricity and electronics; he remembers reading the US Navy manuals on both subjects, in addition to Hermann von Helmholtz's *On the Sensations of Tone*, and becoming addicted to aspirin in the process since the concepts were so difficult to understand.[42] Yet the content of Subotnick and Buchla's negotiations remains unclear, as neither has ever described in detail what they discussed. After a long series of negotiations, a fully assembled and operational Buchla Box simply arrived at the center one day in late 1965, bewildering its recipient, the studio's technician William Maginnis, with its interface, which resembled a space-age control panel, replete with knobs, switches, buttons, indicator lights, and three different types of color-coded cables and jacks (see figure 2).[43]

The two cabinets of the Buchla Box were each divided into three horizontal rows containing two to five individual modules measuring 7 inches tall and some multiple of 4.25 inches wide. Following contemporaneous ergonomic design philosophy, each module had standardized, military-grade knobs, indicator lights, switches, and jacks, as well as professionally embossed front panels with model numbers and a maker's mark in a small sans-serif font that read "San Francisco Tape Music Center, Inc."[44] Each cabinet stood roughly at eye level for a seated user; the size, travel, and feel of every knob, switch, and indicator was finely crafted, and components were mounted to the panel for extra ruggedness. The cabinets were complemented with two controllers that lay flat, each with an array of ten or twelve metal pads, black control knobs, and black and red output jacks. With its control surfaces and patching apparatus standardized, Buchla's modules flattened and abstracted each module's functionality, no matter what was behind the front panel—be it a single, hand-soldered circuit board or a large spring-reverb unit that stretched the width of the cabinet. Buchla's box presented its user with a system of highly specific individual modules, forming a kind of blank slate of choices for the user within a well-defined system, patchable into hundreds or thousands of possible configurations. In Subotnick's words, Buchla had "designed an entire analogue computer-like machine."[45]

Yet the Buchla Box was not quite a musical computer in the way Subotnick had imagined. Rather than "program" it to produce a compositional work such as *Le Marteau* using parameters such as pitch, rhythm, or timbre, its user would configure its modules to perform real-time operations of electronic processes: the use of oscillators to generate electronic sound waves and to modify the amplitude and frequency of other oscillators, both within and outside of the audio frequency range; the start-stop activation of tape transports; the mixing of signals from multiple sound sources; and most importantly, the automatic control of each module, not only by the human user but by the system itself. To create a system that could be programmed to control itself, Buchla separated the electronic signals that traveled between modules into three types: audio signal, which used .141-inch "tini-jax" cables and connectors and could carry the signal of both electronic oscillators and external sound sources; control voltage signal, which used black banana-plug connectors and could either offset or completely control numerous parameters on most modules; and pulse signal, which used red banana-plug connectors and could start or stop the operation of various functions.[46] Buchla's differentiation between audio signal and two distinct types of control signal suggested that this was not a system designed simply to produce new sounds; it was also a system designed to produce new flows of agency and control.

The modularity of Buchla's system emerged from the ongoing discourse within the audio engineering community about the standardization and systematization of separate circuits into a unified modular system. In a 1961 article in the *Journal of the Audio Engineering Society*, the German American audio engineer Harald Bode described such a system as "a complex tone and envelope shaping and modifying device, which may be combined as an integral unit with an alternating sound pattern creating apparatus."[47] Bode's article envisioned what he called "unknown electronic music instrument performances"; in his speculative instrument, "all of the modules or sub-assemblies have input and output jacks, and they can be interconnected with each other by patch cords in any desired order. By doing this it is possible to make up new systems with new performances."[48] Bode suggested using a low voltage control signal, made possible by transistors, to interconnect a variable system of modules to create not only new sounds but "new performances," which meant new dynamics of human-instrument interaction. Although using low voltage signals as control signals was not a new idea, Bode's idea of systematizing multiple modules through the use of a single voltage control standard was; in the wake of this article's publication, several new instrumental systems were developed across the globe in the early 1960s, including the Buchla Box for the SFTMC, Robert Moog and Herb Deutch's first modular synthesizer, and Peter Ketoff and John Eaton's SynKet instrument in Rome, among others.

The Buchla Box resembled these other instruments in its physical design, with vertical, modular cabinets connected to tabletop controllers that could be activated by human fingers, but this resemblance belied a fundamental functional difference. Although the Buchla Box was designed to control the specific material contingencies of the SFTMC, its modular format meant that it could be used for a theoretically infinite variety of purposes, including the control of tape transports, the mixing and modification of signals from external instruments, and even the control of lights, rather than only the "synthesis" of new sounds. Contrastingly, other contemporaneous voltage-controlled electronic musical instruments produced and controlled sound in a way that hewed closely to extant models of musicality. Robert Moog, to whom Buchla has often been compared, designed instruments intended to synthesize sound from electronic sources, using a subtractive "source-filter" model of communication widely adopted by both information theorists and acousticians by the 1960s. This model begins with harmonically rich waveforms or noise and filters it to produce a sonic spectrum that resembles extant "natural" sounds (most commonly the human voice.)[49] Filtering was also a concept that worked not only physically, but also electronically; low-pass, band-pass, and high-pass filters formed the basis for most modern electronic communication systems. Because of the commonality of the filtering metaphor, it was easy for instrument designers to adapt paradigms from electrical engineering to musical instrument building; as Roland Wittje has shown, in the wake of Hermann von Helmholtz's foundational work in the nineteenth century, electrical engineers and acousticians often held a functional equivalency between waveforms, whether purely electrical or transduced into sound to be heard by human ears.[50]

Buchla has described this design philosophy, widely used by other mid-century electronic musical instrument designers, as "designing from the inside out": beginning with a standard function from the domain of electrical engineering, such as filtering, and designing a module that exploits this engineering function for use in music. But Buchla claims that he took the opposite approach, designing "from the outside in": beginning with a desired musical function and then engineering an electrical solution in the form of a module.[51] The modules in the system that Buchla delivered to the SFTMC in 1965 initially controlled the functions of the studio that had been difficult or impossible for its members to control in real time, but as the studio's musicians began to interact with this Buchla Box, they began to develop new ideas for functions that could only have emerged in the moment of performative exploration of the system. Rather than behave like an instrument that would synthesize novel sounds for its human users, the Buchla Box produced unforeseen, nonlinear behaviors both at the level of individual modules and at the level of the system as a whole. Once patched to a certain level of complexity, the system seemed to escape human control, creating new possibilities for real-time cybernetic human-instrument interaction.

MODULES FOR HUMAN CONTROL:
TOUCH CONTROLLED VOLTAGE SOURCES

The idiosyncrasies of Buchla's designs have led both musicians and scholars to brand Buchla as a heterodox engineer. Trevor Pinch and Frank Trocco, in particular, have contrasted Buchla's designs with those of Robert Moog in order to develop a binary comparison between East Coast and West Coast design paradigms. Chief among the differences between these two coasts was the Buchla Box's lack of a standard black-and-white musical keyboard; indeed, they quote Jon Weiss (aka "the man from Moog"), a Moog product specialist, who characterized Buchla as having a "distaste for the [black-and-white] keyboard."[52] But the lack of a musical keyboard on Buchla's first system was not the result of distaste. Rather, Buchla simply saw the use of a black-and-white keyboard as a vestigial holdover from previous types of acoustic musical instruments. With his Box, "[T]he source of sound was electronic rather than vibrating strings, membranes and columns of wind. And to me that meant that it was potentially a new source and therefore instruments based on it would be probably new and different. I saw no reason to borrow from a keyboard, which is a device invented to throw hammers at strings, later on operating switches for electronic organs and so on. To me a keyboard is just a way, a nice way, of dealing with harmonic music, polyphonic, harmonic twelve-tone music."[53]

For Buchla, the modes of musical production made possible at the SFTMC did not necessarily involve "harmonic twelve-tone music," but instead involved the complex routing of dynamic signals that could be processed, fed back or forward, regenerated, run through different media, and eventually sensed by human ears, eyes, or skin. Instead of using black-and-white keyboards to control his system, Buchla designed two different "Touch Controlled Voltage Sources" that could generate both pulsed and steady-state voltage control signals to control any other module that accepted voltage control. One of the controllers (Model 112) had twelve touch plates; the other (Model 114) had ten (see figure 3).[54] These controllers tell the stories of separate engineering problems with solutions that also generated new affordances distinct from the problems they were designed to solve. They also bear the traces of the history of manual/digital interfaces between the human mind and sound, bringing the discourses of deafness,

Figure 3. Model 112 and Model 114 Touch Controlled Voltage Sources. Photo courtesy Andrew Northrop/The MEMS Project.

cyborg prosthesis, and space-age communications to bear on a new systematic paradigm of music as communication. Buchla described each of his modules as "solving a problem," but in many cases, these problems only became clarified once the solution was at hand.[55] It can be instructive, then, to work backward from the solution, as presented in the design of a module's front panel, before exploring the new problem-solution dynamics that Buchla's designs created.

The Model 112 Touch Controlled Voltage Source solved a problem that Buchla noticed early on with studio users who wanted to use a single oscillator as a sound-generating device in the same manner as a tuned string or pipes: audio oscillators could sweep through frequencies with their control knobs, but they could not be discreetly stepped from frequency to frequency. Buchla recalls seeing an early, cumbersome solution for this problem that prevented oscillators from being used in real time, which meant that they could not become the kinds of cybernetic partners in human-instrument interaction that Buchla desired:

> I visited one day at the Princeton laboratory [CPEMC]. [T]hey had reels and reels of tape across the top, and if you wanted half a second of E flat, you'd chop off a section of E flat. It'd be about that long, and they did this thing called intricate taping and assemblage of little pieces of tape and made quite a few compositions from this extra tape that was already pre-recorded at certain very precise frequencies. These pieces were very much in tune because they used laboratory sine wave oscillators to make a tape full of C sharp or something. It was amazing.[56]

To allow a single oscillator to be "played" in real time with stepped frequencies, Buchla designed the frequency parameter of his oscillators to allow for external voltage control and paired them with the Model 112 Touch Controlled Voltage Source, an external module that sat on a table and featured twelve evenly spaced metal pads. When a user placed their finger on one of the pads, the Model 112 would output four discrete control signals: two steady-state voltages controlled by knobs immediately above their respective keys, a variable voltage signal controlled by the amount of skin that made contact with the pad, and a universal pulse signal that would be sent when any key was touched. The two steady-state voltages could be used to control the frequencies of oscillators (possibly one modulating the other, as discussed later); the variable signal could control the amount of modulation; and the pulse could be sent to a voltage envelope generator, which would vary the amplitude of the oscillator's signal to produce a note-like sonic event. Each of the discrete voltage outputs featured two parallel output jacks, allowing the controller to be connected to two independent oscillator modules, or, as described later, the two mirrored oscillator sections of one of Buchla's "Dual Oscillator" modules.

With its twelve keys, the Model 112 could theoretically be configured to play the twelve chromatic pitches of the equally tempered scale—or any other scale—with the Box's corresponding oscillators.[57] In addition, the presence of the variable control voltage determined by finger pressure suggested a similarity to the touch sensitivity of a piano keyboard. But in practice, the Model 112 could not reliably be used to play an oscillator in twelve-tone equal temperament. One major problem was that the oscillators themselves were not stable; their frequencies drifted with minute changes in temperature and fluctuations in the system's power source. Another was that the Model 112's potentiometer knobs were small, had limited ranges of travel, and were neither calibrated to each other nor able to be calibrated by their users; because of this, they lacked both accuracy and precision. A final challenge resulted from the design of the metal touchplates, which behaved differently with different human users based on the moisture content and thickness of their skin, in addition to the ambient air humidity, and even the material of the soles of their shoes that grounded (or insulated) them from the earth.[58] While the impetus for the Model 112's design emerged from the problem of not being able to step through discrete frequencies on a single oscillator, its open configuration and idiosyncrasies generated emergent problem-solution complexes that could not have been anticipated in a design philosophy in which the instrument acted as a tool to execute the will of its human operator.

Like the Model 112, the Model 114 was also intended to solve a specific problem articulated by the SFTMC's users: the inability to control multiple tape playback mechanisms in real time. This was the problem that had motivated Callahan's mechanical relay organ keyboard, Sender's attempted modification of the Chamberlin Music Master, and Maginnis's telephone stepping relay, none of which worked. According to Sender, Buchla's Model 114 was explicitly designed to control a Viking tape cartridge machine that could play ten separate tape cartridges with ten separate playback heads.[59] Such proprietary cartridges held endless tape loops and were used in broadcasting and commercial public address systems; Sender had acquired one at some point in 1964 and had already modified it to utilize its multiple tape heads to play back prerecorded tape loops to create a tape work in his *World Food* series.[60] Yet Sender's modification, which involved turning the unit on its side and hanging tape loops on empty tape reels to tension them with gravity while he manually touched the tape heads to different lengths of tape, was precarious and unreliable.

Buchla's Model 114 featured ten evenly spaced metal pads, each of which activated a discrete pulse signal and a single control voltage signal; the control voltage signal also featured a control for its "decay time," which meant that after the signal was activated, it would decay from its maximum to minimum value in a specified amount of time. Buchla likely designed this controller to work with the Model 107 Voltage Controlled Mixer, which featured ten control voltage inputs controlling the amplitude of ten audio inputs; the control voltage signal would control the amplitude of ten different endless tape loops, and the "decay time" would essentially control the duration of the sonic event before the sound faded out. The Model 114 effectively put ten tape transports at the fingertips of a single human performer, making it possible to use the SFTMC's Viking tape cartridge machine as an instrument for live musical performance.[61] Freed from the burden of splicing tape, the Buchla Box's human user could decide when to activate a tape loop, how loud it would be, and how long it would sound before the signal faded out.

MODULES FOR SELF-CONTROL: DUAL OSCILLATORS AND SEQUENTIAL VOLTAGE SOURCES

With the Model 112 and Model 114 Touch Controlled Voltage Sources, a human user's touch would activate control voltage signals that could initiate events and specify parameters such as frequency, modulation, amplitude, decay time, and even possibly tape speed. But all of these parameters, and the operations to which they were mapped, could also be controlled and activated by a very different kind of user: the Buchla Box itself. This cybernetic self-control could be achieved both within individual modules and also the system as a whole; because most control voltage, pulse, and audio signals were output through two parallel jacks on most modules, it was easy for users to patch the output of one module, or even one section of one module, to control either another module or the same module itself. One example of this module-level self-control was in Buchla's oscillators, of which there were two types: the Model 144 Dual Square Wave Generator and the Model 158 Dual Sine-Sawtooth Wave Generator (see figure 4). The "dual" in their name specified that each module contained two oscillators with a mirrored set of knobs and parameters on a single panel; a user could configure an oscillator module such that one of its oscillators controlled the frequency or amplitude modulation of the oscillator

Figure 4. Model 144 Dual Square Wave Generator and Model 158 Dual Sine-Sawtooth Wave Generator. Photo courtesy Andrew Northrop/The MEMS Project.

on the other side. The output of that modulated oscillator could further be repatched into the input of the first, resulting in cross-modulation with chaotic feedback. The range of these oscillators' frequency parameters seems to have anticipated this; on both models, the lowest specified frequency was 5 Hz, which would not be audible but would be useful for modulation.[62]

Another source of self-control within Buchla's system came from the cheekily numbered Model 123 Sequential Voltage Source (see figure 5), which is described in a brochure that Buchla produced a year after its development: "[The Model 123] produces a sequence of two to eight preselected voltages at each of three outputs. Switching from one voltage to the

Figure 5. Model 123 Sequential Voltage Source. Photo courtesy Andrew Northrop/ The MEMS Project.

next is accomplished by applying a pulse. Indicator lamps show which of the 24 potentiometers are in control. Eight pulse outputs are energized as corresponding outputs are switched."[63]

With a companion Model 146 Sequential Voltage Source that doubled the maximum possible number of preselected voltage steps from eight to sixteen, these modules were designed explicitly to facilitate the production of live electronic music by "sequencing" events in real time. The pulse signal sent out by each step of the sequencer could trigger events, such as the generation of a voltage envelope to control the amplitude of a tape on the Viking tape cartridge machine or the amplitude of one of the system's oscillators; the three control voltage signals assigned to each step could also control the discrete functions of tape machines or oscillators, such as tape speed, frequency, wave shape, or modulation. This would eliminate the cumbersome process of constructing "electronic music" by physically splicing together segments of tape, as Buchla had witnessed at the CPEMC. As Buchla recalled to Joel Chadabe, "My first idea was to reduce the labor in splicing tapes, and that was where the sequencer came from, so if I built a sixteen-stage sequencer, I could eliminate sixteen splices."[64]

But in addition to eliminating the need for splicing tape, Buchla's sequential voltage sources could also be configured to trigger themselves in an endless cascade of control signals, since the pulse signals generated by any one of the sequencer's steps could be sent to start or stop another sequencer, and vice versa. Indeed, this kind of cybernetic self-control was programmed into the very first patch made with the Buchla Box upon its arrival at the SFTMC in 1965. Maginnis, who was there to receive it, remembers staying up until the early morning, programming the Model 123 to control the frequency of an oscillator to produce the first eight pitches of "Yankee Doodle Dandy." Maginnis realized that the Model 123 could make its own "loop" if he programmed the final step of the sequencer to trigger its first step. Leaving the studio that night, he simply shut off the main power switch, since he didn't know how to shut off the Buchla Box itself. The next morning he received an irate phone call from Pauline Oliveros, who had presumably turned on the power to the studio, demanding that Maginnis "make it stop!"[65]

FIRST FLIGHTS: WILLIAM MAGINNIS, 1965

Maginnis's performances with the Buchla Box in November 1965, which are the earliest extant interactions with the system recorded to tape, provide some insight into how new dynamics of human-instrument interaction emerged through a mixture of predetermined programming and real-time interaction. Perhaps because the system had just been delivered and was not yet integrated into the studio, Maginnis's performances do not explore the Buchla Box's ability to control tape machines; instead, they explore the box's ability to control itself. On a reel entitled "Flight," Maginnis utilizes both a touch-controlled voltage source and a preprogrammed sequential voltage source to control a series of square wave and sine/sawtooth wave oscillators, along with other modules such as the Model 160 White Noise

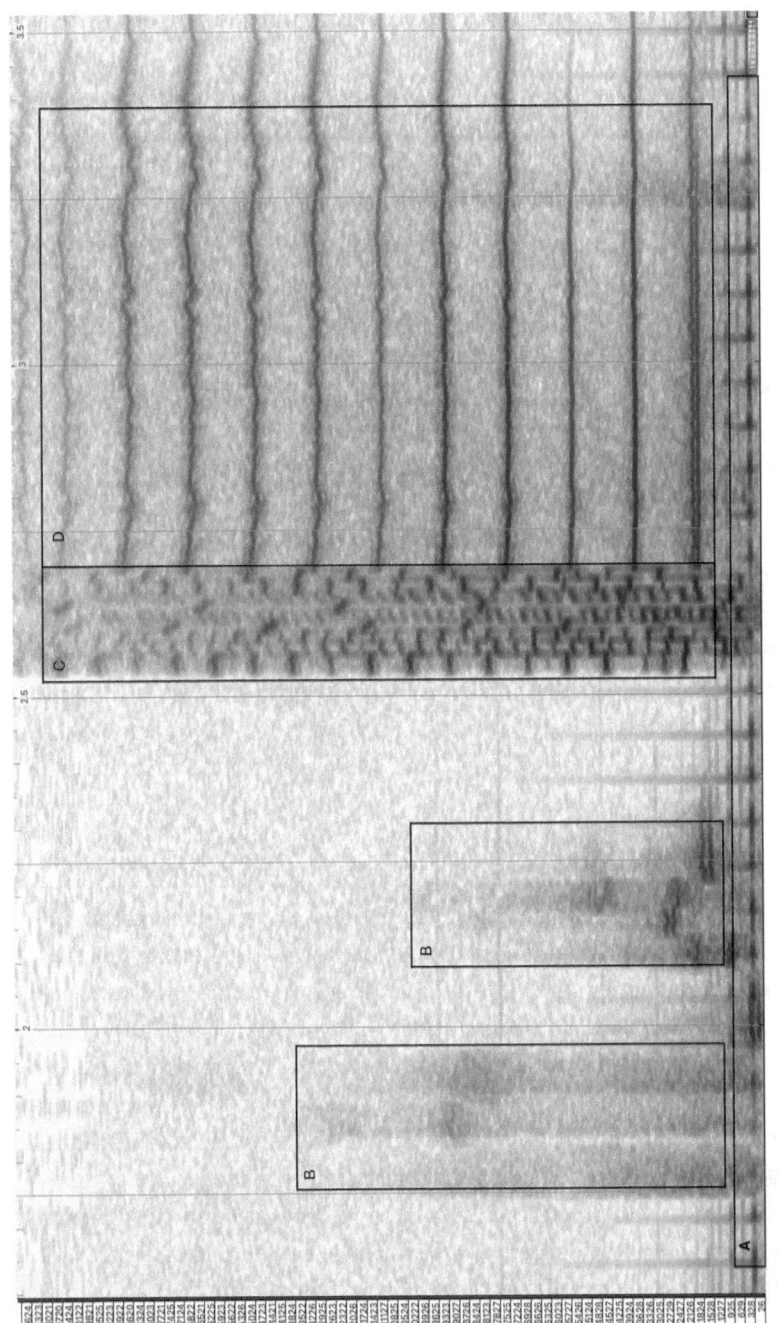

Figure 6. Spectrogram of Bill Maginnis's "Flight" (1965). (A) amplitude-modulated sine-saw wave oscillator signal with additional slow, pulsating amplitude envelope; (B) filtered white noise bursts; (C) extremely fast, quantized jumps in absolute value of fundamental frequency of square-wave oscillator signal; (D) slight frequency modulation of square-wave oscillator signal, controlled either by instability of oscillator, finger pressure on 112 or 114 TCVS, or other external FM source.

Generator, the Model 180 Dual Attack Generator, and the Model 111 Dual Ring Modulator.[66]

"Flight" begins with a sine wave whose amplitude is modulated by a sawtooth wave—most likely utilizing the Model 111 Ring Modulator—and whose signal is also sent through a voltage-controlled gate, itself controlled by a self-cycling voltage envelope (see figure 6). The result is a sound that contains two distinct pulsations, one at the audio level of the waveform and one in the sub-audio range, producing sonic effects that sound like timbral modulation and tremolo, respectively. The fundamental frequency of this sound and the frequency of its amplitude modulation signal increase slowly, most likely manually controlled by Maginnis himself with his hands on the Buchla's knobs; it pans from left to right, likely also manually controlled with a Model 106 Mixer, which was not voltage controllable.

Two other distinct sounds, however, change so quickly that they would have been impossible to manually control, especially if Maginnis's hands were already occupied. One is the sound of white noise bursts with varying amplitude and duration, likely controlled by one of the system's sequential voltage sources. The timing of these bursts' onset and duration is irregular; this could have been achieved by programming the period and pulse length of the Model 140 Timing Pulse Generator to be controlled externally, most likely by steps of the same sequential voltage source that was being controlled by that module. The other sound is a frequency-modulated waveform whose fundamental frequency and amount of modulation varies rapidly, most likely controlled by sequential voltage sources with differerentially timed pulse rates. The overall effect of combining cybernetic self-control with manual control of certain parametrical values is one of extreme complexity; the rates of change between multiple parameters are vastly different, as some are clocked to the rate of the system's self-generated pulses, and some are simply controlled at Maginnis's will.

Maginnis used the word *programming* to describe his first interactions with the Buchla Box upon its delivery to the SFTMC, and beyond the metaphorical use of this term to describe programming the preset voltage values on the Model 123 and 146 Sequential Voltage Sources, Maginnis was perhaps suggesting a more fundamental resemblance between the Buchla Box and a computational device. Indeed, the Buchla Box's interface morphologically resembled the interface of contemporaneous analog electronic computers, which modeled mathematical functions using continuously varying voltages. On such computers, modular sections of operational amplifiers served to perform mathematical operations, connected by the same banana-plug cables used in Buchla's system; potentiometers were used to set the values for such operations, the same as on Buchla's modules; and the system's output consisted of continuously varying voltage, which could be visualized on an oscilloscope or VU meter—or, in Buchla's system, transduced into sound. On analog electronic computers, parameters could be manually adjusted while programs were being executed in real time to study the behavior of the function with different coefficients, or even different scales or magnitudes, but the configuration of the computer's various modules was always only programmable by its human user.

Buchla had undoubtedly used many analog electronic computers throughout his extensive work in aeronautics engineering in the early 1960s. But despite the fact that the Buchla Box morphologically resembled such a computer, its ability to control itself meant that its operation differed in a fundamental way. If it was a computer, it was a computer that could in some sense program itself in real time. In Maginnis's earliest improvisations with the Buchla, we hear such self-programming being effected at tremendous speed, faster than any human hand or ear could comprehend, in the ways that the sequential voltage sources were used to trigger external modules, themselves, and each other. Other early users of the Buchla Box would also pick up on its resemblance to analogue computers; Subotnick remembered that Buchla "designed an entire analogue computer-like machine."[67] But how the Buchla Box configured circuits of control both from itself and from its human user challenged the notion that this "analogue computer-like machine" was simply a programmable computer that executed a static, set program. By design, programming the Buchla Box involved creating complex, nonlinear circuits of control signal between both the Box's human user and the Box itself, operationalizing the "looping circuits of energy and information" that so enamored Buchla and his contemporaries in the Californian counterculture.[68]

Although historians such as Bob Gluck have observed that the Buchla Box "received limited compositional use until [Morton] Subotnick moved to New York [in 1966]," in late 1965 and early 1966 Buchla Boxes were being used extensively by other kinds of users for new kinds of exploratory purposes.[69] These early uses of the Box were difficult to parse through extant models of musicality; not precisely compositional nor improvisational, they instead blurred the boundaries of musical agency and control—human, instrumental, systematic, sonic, and otherwise. Although these uses occurred within socially musical milieus, the subjects who interacted with Buchla Boxes did not understand themselves to be "composers." But rather than simply consider these uses to be "noncompositional," as Gluck's narrative might imply, I propose the term *ex-compositional* to specify the extent to which compositional agency sat outside the human subject in these novel uses of such a complex, cybernetic system. As the remainder of this chapter explores, such uses were often explicitly intended to manifest human consciousness through what came to be called "psychedelic" experiences, a neologism from 1957 that denoted the manifestation of the mind.[70]

I choose the prefix "ex-", rather than "non-," to signify the presence of compositional agency even though it is not wholly situated in the human subject. Contemporaneous with the development of the Buchla Box, Richard Alpert, one of the most popular champions of the psychedelic use of LSD—Buchla's preferred molecule—used this prefix to describe the psychedelic experience as "ex-stasis," of standing beside one's self: "The process of going outside, going beyond learned modes of experience (particularly the learned modes of space-time-verbalization-identity), is called ecstasis. The ecstatic experience. Ex-stasis. The science of ecstatics is the systematic measurement, description, and production of the ecstatic state—that is, the expansion of consciousness."[71]

In an ex-compositional encounter with a technology such as the Buchla Box, the human being goes outside learned modes of experience—not only the learned modes of human-instrument interaction, but also the learned social identities of composer, performer, audience member, technician, engineer, or other roles involved with the activities surrounding musical culture. In such an encounter, consciousness is expanded through exploratory uses of systems that resist legibility and instead offer emergent dynamics that are difficult, or sometimes impossible, to parse. Indeed, the early reception of performances with Buchla Boxes shows just how difficult it was for contemporary musicians, critics, and journalists to understand not only what the Boxes were, but what they were doing. In this way, the Buchla Box functioned as a psychedelic, electronic equivalent to LSD, a drug that came to define the psychedelic experience of many in the San Francisco Bay Area in the exact moment of the Buchla Box's development.

BUCHLA AND THE COUNTERCULTURE: THE TRIPS FESTIVAL

It was not a coincidence that Buchla and his associates began serious explorations of LSD in 1965. In May of that year, a massive quantity of pure LSD was brought to San Francisco by a young clandestine chemist and audio engineer named Augustus Owsley "Bear" Stanley III, who had synthesized the substance in his bathtub in Los Angeles.[72] Through their association with the nascent counterculture, Buchla and many others involved with the SFTMC (notably Sender and Maginnis) had already been experimenting with psychedelics such as mescalin, derived from peyote; the presence of psychedelics at the center formed a social rift that would eventually lead to the center's dissolution by the end of 1966.[73] At its core, this rift illustrated the ideological differences between those for whom the ecstatic, ego-destroying potentials of psychedelic technologies were appealing and those for whom those same potentials threatened their social identity as composers.

Where Buchla, Sender, and Maginnis embraced psychedelics, Subotnick and Oliveros rejected them. Much to their consternation, the SFTMC quickly became a meeting point for what had recently been termed the hippie subculture in San Francisco. Located at the eastern edge of the Haight-Ashbury district, it was also a neighbor to one of the earliest "head shops" in the area, The Magic Theater for Madmen Only, which sold "Victorian knickknacks, pipes, cigarette papers, and weed," and where patrons would often browse after performances.[74] The SFTMC was also an origin point for the "liquid light show," a popular visual performance involving colored liquids and overhead projections, which had been independently developed both by the artist Elias Romero, who performed one at the center in 1964, and the center's own visual director, Anthony Martin.[75] Because of its proximity to this emerging countercultural social world, the studio also began to see an influx of idiosyncratic users, including a man named Peter De Blanc, who was briefly employed as a "member-technician" for a period in May 1965.[76] Blanc, in addition to being a lover of the singer Janis Joplin, promised that he could create "unlimited gain" amplifiers using his own home-cooked schematics, and on one occasion cut every single power cable in the studio to free it from "government electricity" before, in Sender's memory, disappearing into the desert in a Volkswagen Beetle armed with machine guns.[77]

The hippies' arrival coincided with Subotnick's departure. In April 1965 he met Norman Lloyd, a young grant officer from the Rockefeller Foundation (RF), at a performance of one of his compositions in Portland, Oregon; soon thereafter he met another RF grant officer, Boyd Compton, with whom he began discussing the possibility of funding the SFTMC in May 1965. Beginning in the summer of 1965, Subotnick began traveling back and forth between San Francisco and New York City, where he had been appointed the music director of the new Lincoln Center Repertory Theater in the newly constructed Vivian Beaumont Theater, which would open in October of that year.[78] In fall 1965 Compton left the RF to help administer the founding of New York University's new Intermedia program, which hired Subotnick (along with SFTMC visual director Anthony Martin) to begin teaching in the fall 1966 semester. In early 1966 Subotnick permanently moved to New York, where in June 1966 he began creating musical material for his album *Silver Apples of the Moon*, which was eventually released in July 1967 on Nonesuch Records.[79]

But back in San Francisco, before Subotnick used his own Buchla system to create what he called "chamber music 20th-century style," the Buchla Box would be put to use for radically different kinds of musical performance. In late 1965, as Buchla's first system was likely delivered to the SFTMC, these performances began to take place in large venues in San Francisco, with psychedelic rock music, light shows (explicitly inspired by, and sometimes performed by, the SFTMC's visual director Anthony Martin), and freely distributed LSD. The most prominent of these events in the fall of 1965 were organized by Maginnis's friend Chet Helms, a rock promoter who had founded a communal house and production company called the Family Dog, and who would later help establish the band Big Brother and the Holding Company and play a major part in the 1967 "summer of love."[80] Helms began promoting events at the Longshoremen's Hall, a newly constructed modernist concrete octagon that featured a gigantic interior space; Helms's concerts in late 1965 are regarded by many as

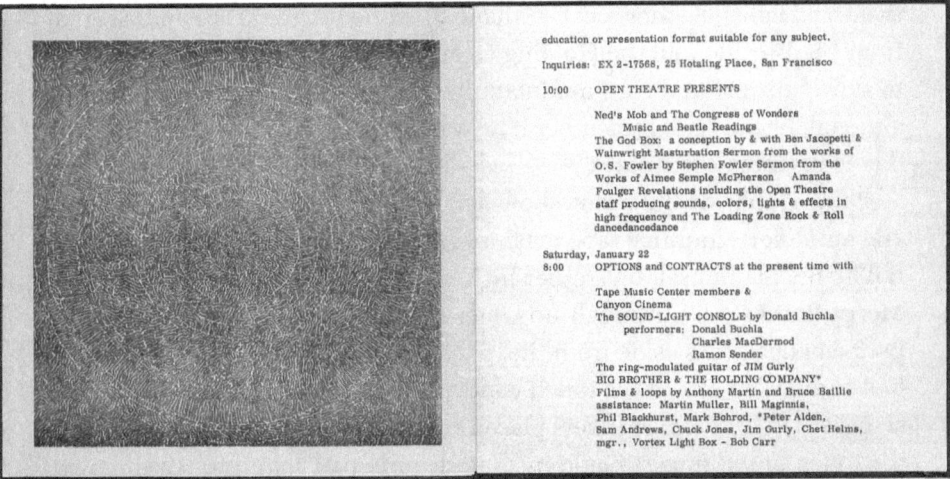

Figure 7. Excerpt from Trips Festival program, January 21–23, 1966, as distributed at the event.

an origin point for San Francisco's psychedelic rock scene.[81] In this same period, Anthony Martin introduced Ramon Sender to Stewart Brand, a young photographer with a traveling "multimedia" show called "America Needs Indians"; Brand asked Sender, in his capacity as the managing director of the SFTMC, to provide technical support for a weekend-long festival at the Longshoreman's Hall that would attempt to bring together all of the nascent hippie groups in San Francisco, which was to be called the Trips Festival (see figure 7).

Many members of the SFTMC were involved with planning and operating the Trips Festival, which took place on January 21–23, 1966. The evening of Saturday, January 22, began with a segment entirely devoted to SFTMC programming; the program lists the "SOUND-LIGHT CONSOLE by Donald Buchla," performed by Buchla, the poet Charles MacDermed, and Ramon Sender; the "ring-modulated guitar of JIM Gurly," and Big Brother and the Holding Company, which Maginnis characterized as "the San Francisco Tape Music Center Band."[82] Throughout the evening, Maginnis lay underneath the venue's stage, high on LSD, with his hand on the main circuit breaker so that he could reconnect it if it broke under the increased electrical demand of the Buchla Box, guitar amplifiers, and theatrical lights. Sender recalled that he planned to route the audio signal from Big Brother and the Holding Company through the Buchla Box and to slowly distort and ring-modulate the signal until it was unintelligible, although he does not remember if he was actually able to execute this plan due to his intoxicated state.[83]

Photographs of the event show a Buchla system on a table with several amplifiers, multiple tape machines, and a rat's nest of cables, manned not by Buchla himself but rather by Ken Babbs, a member of Ken Kesey's Merry Pranksters. Although no concrete documentation of this specific performance exists aside from this photograph, Babbs recalls that Buchla had configured one of his touch controlled voltage sources to send audio signal to ten different speakers placed around the octagonal hall, such that the touch of ten fingers could instantaneously pan sound to ten different locations.[84] This spatialized performance was described in the memoirist Tom Wolfe's *The Electric Kool-Aid Acid Test*, in which he wrote that "the music suddenly submerges the room from a million speakers ... a soprano tornado of it ... all-electric, plus the Buchla electronic music machine screaming like a logical lunatic."[85] Wolfe's memory suggests that the Buchla Box produced sound itself in addition to controlling it, although no audio recording exists of this specific performance.

Despite the relative modesty of these debut public performances, the Buchla Box became the subject of broad adulation in the local press, with extremely optimistic, generalized claims about what the instrument promised to do. Although these descriptions of the Buchla Box compared it to a keyboard, their authors made sure to let their readers know that the box also had something to do with contemporary theorizations of "media," and with the burgeoning information age. The first of these journalistic descriptions appeared in the *Chronicle* even before the Trips Festival, in a preview on January 18; the critic Lou Gottlieb wrote about "a young man who will be there by the name of Don Buchla who will shortly make a million dollars if he decides to mass produce a musical instrument he has invented. It is a keyboard which is connected to an enormous range of the sounds of our time. Every composer should come to see, hear, and possibly play the thing just to know what resources Buchla has made available. Every rock-and-roll record producer will get one before the competition does."[86]

Four days after the Trips Festival, the Buchla Box was used again in a performance by Sender at the Open Theater, a space run by the artists Ben and Rain Jacopetti in Berkeley, where Sender had recently moved; a listing in the *Chronicle* describes the Buchla box as an "experimental sound and light console," and Sender's performance as an "audio-optic program," emphasizing the generalized sensory experience of the performance.[87] A subsequent review from January 29, 1966, although unfavorable, asserted that the Buchla Box "allows the electronic musician direct access to the sound medium, without resort to the intervening IBM program card. It could be the precursor of fascination [*sic*] developments in the field of electronic music."[88] The Buchla Box was parsed by these journalists as something between a musical keyboard, a "console," and a computer; whatever it was, its effects on the sensory apparatuses of humans around its performance was radical, escaping easy description. And in the months after these initial performances, Buchla Boxes were used in ways that would further explore their potentials to effect changes in sensation and human consciousness, through their continued use in psychedelic, ecstatic performances throughout the Bay Area.

EX-COMPOSITIONAL USES OF THE BUCHLA BOX: KEN KESEY, THE AAA, AND THE HELL'S ANGELS

Immediately after the Trips Festival, several attendees asked Buchla to construct boxes for various uses that involved neither the composition nor the performance of musical works, but instead involved the real-time utilization of sonic and electronic signals to explore psychedelic, mind-manifesting performances. Ken Kesey, the leader of the Merry Pranksters, asked Buchla to build a customized box for his traveling school bus, *Furthur*, on which he had already installed a "time lag" system consisting of multiple hardwired microphones, tape machines, headsets, and loudspeakers placed inside and outside the bus, which would record sound and play it back with a variable time delay.[89] The *Furthur* bus had made its maiden voyage in 1964, and Kesey likely wanted to refurbish and expand the bus's functionalities by 1966. For Kesey, Buchla produced a small box containing four modules: a two-channel microphone preamplifier, a six-channel

Figure 8. Ken Kesey's Buchla Box. From left to right: Dual Reverberator Model 190; Dual Microphone Preamplifier Model 170; Six Channel Mixer Model 106; a custom amplifier module. Courtesy of the National Music Centre, Calgary, AB.

audio mixer, a two-channel reverb module, and a power amplifier with two stereo headphone outputs and one set of stereo speaker outputs (see figure 8). Although this system did not contain modules that utilized control voltage, it simplified and streamlined the extant "time lag" system, putting its control into a single unit and adding to its combinatorial possibilities for routing sonic signal. Significantly, because Buchla's modules featured dual paralleled outputs, both the dry and reverberated signals of the microphones could be mixed together in different amounts, sent to different tape machines, and routed to the left and right stereo channels of both headsets and loudspeakers throughout the bus, then be fed back into the bus's extant tape machines, permitting the configuration of several cybernetic, nonlinear feedback systems. The Buchla Box on *Furthur* increased the order of magnitude of possible interconnections between input and output, thus behaving like a psychedelic technology for the expansion of sonic consciousness through real-time cybernetic feedback.[90]

Another prominent Trips Festival attendee who commissioned Buchla to build a system in mid-1966 was the psychedelics researcher Richard Alpert, who planned to donate a Buchla Box—along with 100,000 μg of

LSD—to a nascent communal rock band called the Anonymous Artists of America (AAA).[91] Alpert's commission for the AAA was among the earliest orders Buchla received, after an early 1966 commission by Subotnick for a second system in his New York studio and an order from the LSD chemist Owsley Stanley, discussed later.[92] By pairing the Buchla Box with LSD, Alpert drew an explicit comparison between their respective mind-manifesting potentials as psychedelic technologies. Indeed, Alpert conceptualized the psychedelic experience using computational metaphors, which had also characterized Buchla's conception of his Box. In a report on his activities at the International Federation for Internal Freedom, a short-lived intentional psychedelic community that Alpert ran with Timothy Leary and Ralph Metzner from a hotel in Zihuataneho, Mexico, Alpert described the psychedelic experience in explicitly computational terms:

> The central metaphor is as follows. The human brain contains over ten billion cells. Any single cell can be in interconnection with up to twenty-five thousand other cells. About one billion impulses flood into the cortical computer each second. The potentialities of consciousness at any one second are thus seen to be on the order of $(1,000,000,000) \times (10,000,000,000)^{25,000}$. The educated adult utilizes about five thousand concepts to experience the world within and without. An astonishing filtering and constricting process occurs which reduces the enormous potentials of consciousness to the few cultural modes of experience routinely employed. Psychedelic drugs are seen as interfering or counteracting these reductive processes so that subjects are able to experience immediately, beyond the limits of the learned cultural programs.[93]

Alpert conceptualized the human mind as an organic electronic computer through which the signals of sensory information could flow through ten billion possible connection points; psychedelic drugs enabled humans to transcend "learned cultural programs" by allowing those ten billion possible connection points to be reconnected in new and unknown ways. Achieving this state of "ecstasis," as Alpert called it, required the real-time exploration of one's own consciousness through various techniques including "fasting, contemplative focus of attention, optical alterations, yoga exercises, sensory deprivation, and the ingestion of foods and drugs."[94] Although not named explicitly in this list, experimenting with the Buchla Box involved routing a human's gestural "output" and sensory "input" through a system that was itself modeled after an analog computer, with hundreds of possible connection points that could be interconnected in thousands of ways; its use could "serve as [a key] to free the individual to experience new states of consciousness."[95]

The AAA, for whom Alpert commissioned his Buchla Box, coalesced around a young, charismatic, gay Stanford undergraduate named Lars Kampman, whom Alpert knew socially through the Perry Lane bohemian enclave in Menlo Park.[96] In June 1966, inspired by attending events in San Francisco such as the Trips Festival earlier that year, Kampman and his friends decided to form the AAA; with an inheritance of $20,000 from his family's lumber fortune, Kampman bought a full set of instruments for the band and began housing them in his cottage on Homer Lane, adjacent to Perry Lane. Soon thereafter the band relocated to a larger property called Rancho Diablo in the Santa Cruz Mountains, where Buchla delivered their system a few days before their first public performance in July or August 1966 at the wedding of Merry Pranksters Lee Quarnstrom and Space Daisy at the Filmore in San Francisco.[97] Although no recording exists from that performance, Quarnstrom remembered the AAA as "a band built around a Buchla machine, a music synthesizer built by a genius Don Buchla."[98]

Although the Buchla Box symbolized the core of the AAA's identity to Quarnstrom, Kampman remembers that it did not integrate into the band's attempts to produce rock music. Instead of serving a musical function, it served an ecstatic, psychedelic function for its human operator: "[I]t wasn't clear to anyone how it would function in the band. [. . .] It never became clear. For the time being, it made desultory shooshing and swooping sounds at frequencies that seemed completely random."[99] In Kampman's memory, the AAA's Buchla Box forced all entities involved—its user, itself, and its sound—to remain radically ambiguous. Although this ambiguity did not sound like rock music, it did resonate with the personal philosophy of the Buchla player in the band, a young University of Chicago dropout named Len Frazer. Because of his early departure for the Bay Area from college, Frazer's coursework at Chicago consisted entirely of the philosopher Richard McKeon's core curriculum in "Ideas and Methods," in which McKeon taught a philosophy he called "operationalism": "Meanings are to be developed out of the *relations* and distinctions between and among the terms" schematized as variables in matrices.[100] This structuralist philosophy, deeply informed by systems thinking, promised a rigorous exploration of the combinatorial possibilities of meaning, valorizing ambiguity as "the source not only of philosophic problems but also of discussion and progress."[101]

Frazer's use of the Buchla Box with the AAA enacted this philosophy, seeking out new meanings through the systematic and schematic questioning of all inherited variables. The "shooshing and swooping sounds" that Kampman found antithetical to rock music were radically ambiguous, resulting from a continuous exploration of the combinatorial possibilities of the flows of signal between the Buchla Box's modules. Kampman's additional complaint that the instrument could not be tuned furthered the instrument's resonances with McKeon's operationalism; rather than allow for a "single, true definition of any of our concepts" such as a definition of music as adhering to twelve-tone equal temperament, the instrument provided an arbitrarily defined matrix of possibilities from which an endless process of discovery could emerge.[102] In this sense, Frazer's use of the Buchla Box positioned the box as an ex-compositional partner in an ecstatic psychedelic event, reconfiguring the possible meanings of sound in real time (see figure 9).[103] Though Frazer worked closely with the Buchla Box for over a decade, its centricity in the AAA was perhaps what kept them from ever becoming commercially successful; the band's output was limited to live performances and a brief cameo in the 1968 hippie exploitation film *Psych-Out*.[104] Although Frazer and the AAA worked with technologies such as LSD and the Buchla Box to manifest their minds, these interactions did not produce artifacts legible as musical works.[105]

The AAA's close association with LSD was shared by Buchla himself, primarily through his long friendship with Stanley, who had become the chief supplier of the drug to the countercultural scene in the Bay Area, and who also became the sound technician for the Grateful Dead soon after the Trips Festival, in February 1966. Stanley supplied Buchla with LSD, and in return, Buchla constructed a customized Buchla Box for Stanley with cherry-red front panels sometime in 1966. These panels were rumored to have been dipped in Stanley's "White Lightning" LSD, such that the box's operator could simply lick their finger after touching a panel to receive a dose. Although many experts have long doubted this lore, in 2019 a television engineer named Eliot Curtis claimed to have experienced a nine-hour

Figure 9. Len Frazer and his Buchla box, 1966, Rancho Diablo, CA. Photograph by Dean Quarnstrom, courtesy Evan Quarnstrom.

psychedelic trip after touching a crystalline substance found embedded in a potentiometer on one such "red panel" module; laboratory tests confirmed that the substance was 99 percent pure LSD.[106]

Through Stanley and the drug trade, Buchla also developed a relationship with the outlaw motorcycle club the Hell's Angels, who had begun to insinuate themselves into San Francisco's psychedelic rock scene after being introduced to Ken Kesey by the gonzo journalist Hunter S. Thompson in August 1965.[107] In 1965 and 1966, Kesey held a series of parties at his house in La Honda in the Santa Cruz Mountains that were attended by "college professors, vagrants, lawyers, students, psychologists and high-style hippies," as well as the Pranksters and the Angels; attendees united in their embrace of LSD above all else, which led to a disturbing dynamic in which the Angels' violent, misogynist behavior was openly tolerated.[108] Buchla later worked with the Hell's Angels directly at the infamous Altamont Speedway Free Festival in 1969, where the gang had been hired to provide "security," resulting in the stabbing death of a fan named Meredith

Hunter; Buchla was running the sound system.[109] In a 2008 interview with Maggi Payne and David Bernstein, Buchla remembered his relationship with the gang, connecting them to Kesey and their mutual interest in LSD:

BUCHLA: And I did a lot of recording with the Hell's Angels.
BERNSTEIN: What did they want?
BUCHLA: They were a little bit associated with Kesey's bus and were involved with drug distribution. [Laughs] It all seemed fine to me. I got some interesting tapes from that source.
PAYNE: They were musicians as well?
BUCHLA: No, you don't have to be a musician to make a tape.[110]

CONCLUSION: WHAT KIND OF PERSON CAN USE A BUCHLA BOX?

Buchla's response to Payne—that one doesn't have to be a musician to make a tape—hints at the challenge that new audio technologies in the 1960s posed to the social identities of their human users. As an engineer trained in information theory and cybernetics, and as a devoted psychonaut, Buchla was interested in the changes that new electronic and chemical technologies could produce within the complex interconnected systems that comprised the human being, conceptualized in informational and computational terms. Buchla explicitly characterized his systems as computers: a 1966 brochure described the system as enabling "live, on-line, real-time performance of electronic music without tape buffering."[111] Though commonplace today, the terms *on-line* and *real-time* were novel in 1966 and were nearly exclusively associated with computing; on-line specified that a human was working immediately with a computer (as opposed to programming it with punch cards), and real-time specified that the computer would output information as it was processed instead of in batches. In addition, Buchla described the performance of prerecorded audiotapes as "tape buffering," which in the domain of computing denoted the asynchronous storage and retrieval of information printed on magnetic tape. With the Buchla Box, the performance of electronic music meant the programming and execution of novel, cybernetic computer programs—both with the electronic computer and the biological computer—in real time.

The imprecision, unpredictability, and emergent nature of such cybernetic programs appealed to Buchla, along with early users of Buchla Boxes such as Maginnis, Kesey, and Frazer of the AAA, because it mirrored an ecstatic, psychedelic exploration of the self through expanding the computational possibilities of human consciousness, just like the ecstatic explorations each had undertaken with psychedelic chemicals. But for composers such as Subotnick, the nonlinearity and imprecision of the Buchla Box made it impossible to use as what he considered to be a musical instrument. In a 2016 conversation with Suzanne Ciani, a composer who began to use Buchla systems in 1968, Subotnick remembered this tension: "This was an argument that Don and I had from day one. He wanted to make a musical instrument. I said, 'This is not a musical instrument. This is, at best, an instrument to make instruments [. . .].' He didn't understand the nature of an instrument. He said, 'I'm going to make musical instruments.' I said, 'Who's going to play them? How are they going to learn to play them?'"[112]

Subotnick's differentiation between a musical instrument and an instrument to make instruments, and his bewilderment at the idea that a performer could learn how to play a Buchla Box as a musical instrument, implies that a musical instrument must in some way respond in a stable, predictable, and precise way to human gesture. But stability, predictability, and precision were extremely difficult to achieve with the Buchla Box. Not only could it be configured in thousands of combinatorial patches; many of those patches would effect cybernetic self-control from within the system itself, veiled by the opacity of the patch's complexity. This reduced the ability of its human user to reliably control the system as a musical instrument in Subotnick's terms. But in Buchla's imagination, his musical instrument was not simply an extension of the human operator's will. It was, as he wrote in a later poem, a "teacher."[113] Indeed, in this 1971 poem, which is the only text authored by Buchla that describes his philosophy of technology, the instrument is characterized as a psychedelic interface between "biocomputer" and "hardware computer," connected in an endless, cybernetic dance:

WHAT THE MACHINE CAN AFFORD
AND PRODUCE IS A NEW LANGUAGE

CULTIVATING AND REFINING
IN THE LISTENER A NEW SENSE
OF PERCEPTUAL MECHANICS WHEN
THE LISTENER IS ALSO THE
OPERATOR-CREATOR HIS VISUAL
AND AUTO-MUSCULAR SENSES ADD
TO HIS INTERPRETIVE MODEL
OF THE UNFOLDING AUDITORY
PROCESS THE OPERATOR
FUNCTIONS IN A SPECIAL
RELATIONSHIP TO HIS OWN
PERCEPTUAL FEEDBACK SYSTEM
 A SYSTEM IN WHICH THE MACHINE
FUNCTIONS AS A CRITICAL
INTERFACE WITHIN THE CIRCULAR
CIRCUIT CRITICAL THRESHOLDS
EXCEEDED ENABLING
BIOCOMPUTER TO ACHIEVE
INTERLOCK WITH HARD-
WARE COMPUTER THE
PROGRAM AS A PATCH-UP
AND SET OF SETTINGS
OF DIALS AND SWITCHES
REPRESENTS A SCORE
BUT A SCORE WHICH IS
VARIABLE GESTURE-
EXPRESSION-MEANING-
LOGIC-STRUCTURE-STATEMENT
ARE ARRIVED AT THROUGH
CYCLES OF INPUT AND OUTPUT[114]

Rather than a musical instrument that produced novel sounds, the Box was imagined as a human-computer interface that enabled both new sounds and new vectors of control of sound to emerge through human-computer interaction. Buchla's poem suggests that unlike the "realization of ideation" that occurs with acoustic instruments, the ideation made possible through his Box was different, since the "automatic processes" with which humans previously created musical ideas could be internally modified with a level of complexity theretofore unknowable and unachievable within the limited parameters of a traditional instrument. The Buchla Box did not extend the human subject; it ecstatically co-constituted it, locating the self outside of the self.

Buchla was not the only person at the SFTMC interested in rethinking what it meant to be a musicking human subject—conceptualized as a complex collection of various biological, perceptual, and cognitive systems from which consciousness emerged. Throughout the same period that Buchla was developing his Box, SFTMC manager Pauline Oliveros simultaneously developed what she would later call "a very unstable nonlinear music-making system" that incorporated not only the basic elements of the center's studio, such as oscillators, tape machines, patch bays, amplifiers, and speakers, but also her own body, its sensory apparatuses, and its fleshly appendages.[115] With this system, Oliveros swerved from questions surrounding compositional agency and psychedelic potential, and instead began to investigate the possibilities such new technology would produce for her libidinal, sensory relationship to sound and improvised performance.

3   The Patchwork Girl

PAULINE OLIVEROS AND
THE CYBERNETIC BODY

On the same evening that Buchla's Box was first being demonstrated at the SFTMC in 1965—so the story goes—Pauline Oliveros was holed up in 321 Divisadero Street's attic studio, staring at oscillators.[1] She wanted to figure out how to use them to make electronic music, but not the kind of electronic music the men downstairs were arguing about. Unlike her colleague Morton Subotnick, Oliveros was not particularly interested in extending a composer's "heroic vision" through a purified domain of electronic sound; unlike Donald Buchla, she was not particularly interested in the psychedelic manifestation of the mind, either. What she shared with Buchla, however, was a curiosity about what electronic technologies could teach her about herself: not only her mind, but also her body, its senses, its desires, and its pleasures. As a musician whose identity as an openly queer woman was marked as different within the nearly entirely androcentric social milieu of American New Music, Oliveros had already developed performance practices with a gynocentric group of nonmusicians to explore what sound could reveal about their bodies' seemingly unconscious inclinations. Staring at the oscillators, she thought about how she might recreate an experience she had enjoyed as a teenager with her accordion—the satisfying buzz of combination and difference tones, which emerged when two adjacent keys were depressed and the bellows was drawn hard, producing a ghostly third perceived pitch as well as slower vibrations that were physically conducted from the accordion into her body.

Staring at the oscillators, she thought that perhaps, if she just figured out how, she could use them to create these tones electronically. Patching two Hewlett-Packard HP 200CD oscillators into amplifiers and speakers, she set their frequency dials above the upper limit of human hearing and swept them past each other, expecting to hear a new, third, ghostly sine wave. Nothing happened. Thinking they were just too quiet, she inserted a line amplifier into each oscillator's signal path, electrically summed the amplified signals in the patch bay, and amplified them some more. She also ran them through a wildly unstable tape delay system, reiterating the slightest movement of her hands at short intervals, mere fractions of a second. This time, eerie, ghostly, heavily distorted sounds emerged in the room. As Oliveros's close friend Stuart Dempster remembers, "In [her early electronic] pieces she'd get the tone generators and then find the 'garbage' that they would do together. This makes it like they're living somehow, in a way that straight electronic music to me never did, for my ear."[2] What was the "garbage" that Oliveros and the oscillators would do together? In what way were they alive? And what, exactly, was not so "straight" about this music?

Dempster's characterization of Oliveros's early electronic work as "garbage" may seem insulting, but in fact it may be an apt description: this work produced tapes that traced Oliveros's experimental, improvised encounters with ad hoc instrumental systems, filling reel after reel with traces of interactions that challenged the notion of a subject in control of their body and of the instruments at their fingertips. Listening to these tapes reveals the illegible excesses that would usually be edited out, relegated to the garbage. Like the other musicians studied in this book—particularly Alvin Lucier and Sun Ra—Oliveros had experiences with technological instruments that held desubjectivizing potentials, which she came to embrace. While Oliveros was trained to believe that the composing subject ostensibly exerts control over musical material through planned actions with musical instruments, her experience with electronic technologies of magnetic tape and oscillators disrupted an understanding of the human subject as a monolithic entity that possesses and controls the body. She found that the subject could not speak for the body; the body spoke for itself. In 2016,

Oliveros summarized her early experimental protocol, describing not only these initial experiments, but indeed much of her career: "This method has continued to serve me well throughout the years: play, record, listen, and then discuss—then play and record some more. We trusted our bodies to deliver the goods."[3]

Although Oliveros began to trust her body, her early practices also questioned the integrity of that body, and particularly highlighted that body's dynamically shifting boundaries. Beginning in the mid-1950s, Oliveros began to conduct experiments both internal and external, fleshly and machinic, first with magnetic tape machines and a variety of improvised, home-made instruments and later with electronic oscillators. These experiments complemented her career as a composer of notated works for traditional instruments, which began at a young age and continued with her studies under Robert Erickson, whose deep investment in contemporary scientific discourses of cybernetics and systems thinking led Oliveros to expand her experiments beyond traditional instruments and compositional methods. Combining these discourses with the instruments she had at hand, Oliveros began to treat herself as an object of study by inserting her physical body into an experimental system of instruments, executing experimental protocols, and listening back to the sonic results. The material traces of these early experiments, inscribed as signal on magnetic audiotapes, revealed the noisiness, instability, and unpredictability of her body. As in the electronic musical practices developed by Alvin Lucier and Sun Ra, the early electronic music of Pauline Oliveros posited a performing body that resisted legibility in extant social and aesthetic worlds of music.

The performative body revealed through these experimental practices was, in Oliveros's words, a patchwork: both the literal patchwork of patching cables into input and output jacks on various pieces of electronic equipment and the metaphorical patchwork of patching her own body into that equipment, through the output of her fingertips and the input of her ears. Oliveros's earliest description of this experimental practice came in a letter written to the composer and pianist David Tudor in April 1965: after describing progress on a traditionally scored theatrical work, *Pieces of Eight*, Oliveros signed off, "Otherwise I'm very busy with sundry research of a patchwork quilt nature, which goes in delightful directions. Come back and we will make a new kind of concert."[4] Describing this work as research, rather than composition, performance, or even improvisation, suggests that working with oscillators was an act of discovery and provisional construction. Rather than expressing herself, she was constructing a self out of many separate parts, like patchwork—the process of constructing a textile out of many separate pieces, threading them together to form a coherent whole. Not only is patchwork a common feature of quilting, an activity commonly associated with femininity, community, and care work; it is also the word used to describe the interconnection of separate instruments in an electronic music studio, creating an experimental system such as the one Oliveros developed with oscillators, amplifiers, tape machines, and speakers. Patchwork became a major metaphor for Oliveros's conceptualization of her early work with electronic audio technologies and continued to inform her writing about her own work well into the 1970s, particularly in a series of articles written for the magazine *NuMus West*, which invoked a character from the children's novelist L. Frank Baum: Scraps, the Patchwork Girl of Oz.[5]

In Baum's story, the Patchwork Girl is created by a crooked magician whose wife wants to have a female servant. This ideal woman—subservient, constructed, without agency—is sewn together from fabric scraps and brought to life by virtue of a magical powder along with the sound of a phonograph playing a military march. Yet in the moment of her birth, the Patchwork Girl takes agency into her own hands: she knocks over the magical life-giving powder, spilling it on the phonograph, which gives it its own consciousness and frees her from the bonds of her male master. The Patchwork Girl is then free to explore the land of Oz, all the while being chased by the phonograph, who insistently plays recordings of a "highly classical composition," the only music it is able to play.

Oliveros identified with the Patchwork Girl, as a woman whose body was inscribed by an androcentric society but who created her own agency, as a woman who herself gave agency to instruments of sound recording, and as a woman who was indeed chased through the compositional world by the legacy of classical music. Like the Patchwork Girl, Oliveros challenged modalities of listening through a fundamental self-determination of the technology of her own body and its relationship with musical instruments. As Oliveros would explore in this early moment in her career, the patchwork of the electronic music studio—patching together instruments, tape machines, signal generators, mixers, microphones, loudspeakers, and more—was analogous to the patchwork construction of her own body, producing an emergent musical subjectivity that was not entirely her own. With these patchwork systems, and through these emergent musical subjectivities, Oliveros developed performance practices in which sound behaved as an informational signal connecting the domains of her acoustic perception, sound waves in air, and electronic instruments. The performative dynamics of these systems emerged in moments of human-instrument interaction and evaded both human control and human understanding; Oliveros's systems effected a cybernetic musical agency in which the human subject was only one part of the patch.

PATCH 1: HUMAN/TAPE RECORDER

One of the earliest instruments with which Oliveros constructed a patchwork self was a magnetic audiotape recorder. In articles, lectures, and interviews throughout her long career, Oliveros often returned to her first encounter with a tape machine, remembering it as the first time she discovered that her awareness of sound could be changed through technology—first through a rupture in the integrity of the human subject, and then through a suturing of that rupture by patching in new parts. The creation of this first patchwork self involved her auditory, sensory, and cognitive apparatuses, a tape machine, a microphone, headphones, a room in an apartment building, and the city of San Francisco. By configuring the microphone and tape machine in a certain state—levels, gain, mic placement, and so forth—and executing an experimental protocol with this system, Oliveros produced a sonic signal that radically altered her consciousness of sound.

As she recalled in 1998: "[In 1953], I placed the microphone of the recorder in the window of my San Francisco apartment and recorded the sound environment. Little did I realize the extent of the impact that this simple act would have on me. Although I thought that I was listening while recording, I was surprised to find sounds on the tape that I had not heard consciously. With this discovery, I gave myself a meditation: 'Listen to everything all the time and remind yourself when you are not listening.'"[6]

This anecdote, and the frequency with which Oliveros told it throughout her life, recalls John Cage's well-known story about his experience in Harvard's anechoic chamber, a room without echoes, in 1951. In that story, Cage entered the chamber, designed by the acoustician Leo Beranek for the National Defense Research Committee, and heard two unexpected sounds; an engineer told him that they were the sounds of his circulatory and nervous systems, which led Cage to the conclusion that "there is no such thing as silence."[7] Oliveros's anecdote was similarly a story about the revelation of otherwise "silent" sounds. But unlike Cage's anecdote, in which the technicity of the anechoic chamber remained transparent, Oliveros's anecdote valorized the technicity of both the tape-recording apparatus and her own body. Oliveros's experience led her to question her body's sensory and attentive capacities through a recursive experimental protocol, incorporating the tape machine into the listening body: first record sound onto tape, then rewind the tape, listen to the tape through headphones, change one's listening, and record again. Oliveros's bodily organs, along with the external instruments of tape recorder and microphone, became part of the same patchwork system, and Oliveros's experimental protocols for exploring this system produced an artifactual signal that endured on tape, evidence of this new body's capabilities.

Another major difference between Cage's and Oliveros's anecdotes was the conclusions they drew from them. Oliveros quickly turned her attention to the internal sonic world of her self through what she called a practice of improvisation—an activity that Cage famously disavowed in favor of chance operations for composition. Oliveros had been taught to improvise by her teacher, Robert Erickson, first at San Francisco State College between 1954 and 1957 and later at the SFCM between 1957 and 1959. Erickson embraced the technique of improvisation in the service of producing compositional material; deeply immersed in structuralist, systems-theoretical philosophies of music, he held that composers could empirically discover latent musical material in themselves by improvising and recording their performances. Erickson's pedagogy embraced biological metaphors for music, joining a long nineteenth-century tradition of organicist metaphors for musical composition, but with a cybernetic twist. Unlike E. T. A. Hoffmann's perception of an "organic structure" from a single "germinating seed" in Beethoven's instrumental music, for example, Erickson saw compositions themselves, just like their human composers, as self-regulating organisms reacting to dynamic environments.[8] In his first published textbook, *Sound Structure in Music*, Erickson describes a musical composition as such a cybernetic system: "Stated bluntly, a musical composition is a system in the same way that an organic entity is a system. This is meant as a simple, non-metaphysical statement. [...] Music maintains its configuration in a constant environment and counters alterations in the environment."[9] Erickson quickly incorporated this cybernetic understanding of music into his pedagogy. If music was a system in the same way that living beings were systems, then empirically exploring those systems through improvisation could yield new insights for the practice of musical composition.

Improvising with Tape

In 1957 Oliveros, along with fellow Erickson students Terry Riley and Loren Rush, formed a group to perform Ericksonian group improvisation. Riley had been commissioned by the sculptor Claire Falkenstein to make a soundtrack for her short film *Polyester Moon*, and he invited Oliveros and Rush to collaborate with him.[10] Meeting in the studio of the radio station KPFA, where Erickson had been the music director since 1954, they executed an experimental protocol of improvising while recording to tape, listening back to the tape, discussing, and improvising again. Oliveros brought her French horn, Rush brought a koto and percussion instruments, and Riley played KPFA's piano; the group would simply begin improvising "without discussing what to do."[11] After completing five cycles of improvisation, feedback, and discussion, the audiotaped artifact of their fifth iterative improvisation was used for the film's soundtrack. These improvisations, recorded and archived by KPFA, are reminiscent of the sparse, gestural, and atonal musical idioms of European avant-garde music; there is never any identifiable pulse or tonal center, and instrumentalists rarely repeat any melodic or harmonic material.[12]

Later in her life, Oliveros cited this early group improvisation as the core of a lifelong empirical practice of self-exploration, characterizing it as an expansion of the institutional compositional practice in which she had been educated and socialized as Erickson's student: "[W]e considered our work to be the first group improvisation stemming from art music of the time. Indeterminacy and open form already provided for some performer choice with limited possibilities offered, but the indeterminate element was kept under strict control by composers such as John Cage, Earle Brown, Lukas Foss, and others. Improvisation was not yet a real option in this music."[13]

Oliveros's characterization of these early improvisations is notable in relation to the other musicians brought together in this book in two specific regards. The first is that Oliveros characterized the musical output of her improvisation with tape as emerging from her body, not her conscious mind; as she later wrote, "the tape recorder enabled me to more deeply access body consciousness through improvisation."[14] Locating musicality in her body rather than her mind starkly differentiated her practice from that of her colleague Morton Subotnick, for whom electronic media was fantasized as a means for immediately extending compositional ideas from the mind. In this sense, Oliveros's embrace of technoscience moved her away from Subotnick's valorization of traditional compositional agency and more toward a distributed, embodied sense of emergent cybernetic agency. But while Oliveros valorized body in addition to mind, she still ultimately thought of improvisation with tape as a technique for emancipating the performing subject by transcending the limitations of indeterminacy, open form, and structured improvisation developed by Cage, Brown, and Foss, respectively. As discussed in chapter 5, valorizing the performer's individual freedom also starkly differentiated Oliveros's cybernetic practices from those of Sun Ra, who claimed that his cybernetic performances were not evidence of the freedom of an individual human subject, but rather were evidence of a fundamentally unknowable self, channeling a higher cosmological order.

Oliveros's cybernetic practice of improvisation, as taught to her by Erickson, sought to both expand the scope of human subjectivity from the mind to the body and to emancipate that human subject through immediate exploratory performance. This was elegantly summarized in an Ericksonian mantra about sound written in a 1960 letter to Oliveros: "I *think* sounds (not *about* sounds)."[15] From this basic premise, Oliveros developed an experimental system for sonifying the process of "thinking sounds." Iterating musical intuition over and over, listening back to sounds unintentionally yet audibly produced, began to beg the question: Who, or what, was performing, sounding? Oliveros began to think seriously about her lungs, mouth, lips, and hands, systematically attached to the horn, as equal partners in a systematically described circuit of musical signals. This circuit was performative and self-regulating; it did not seek to represent itself, but instead, simply to exist in the moment of performance.

Although Erickson ultimately thought that such performances would serve to develop musical material to be used in compositions, Oliveros went one step further in valorizing improvised performance over the scored representation of sounds. In October 1960, Oliveros wrote to Erickson to ask if a musician could be trained to be a "concert performer" if they were only trained to improvise. In his reply, Erickson supported the practice of improvisation as a means toward "security" and "mastery and control" of instruments, but still insisted that proper technique for "concert performances" was necessary as well.[16] Despite her teacher's allegiance to concert music, Oliveros soon began to practice improvisation with tape as a means for a different kind of performance by a different kind of musical self, one that did not maintain mastery and control over instruments (and over itself), but instead sought other models and other motivating questions for musicality.

Moving further away from Erickson's allegiance to the institutions of concert performance—and further from the androcentric world of institutional music in general—Oliveros began to teach her protocol of improvising with magnetic tape machines to a small group of amateur female students, including her romantic partner at the time, the painter Laurel Johnson.[17] Improvisation became a group practice of self-exploration in which sounds would not be evaluated for their compositional merit, but rather simply observed. In a July 1960 interview on the local radio station KPFA, Oliveros described the experience of listening to her group's tape-recorded improvisations: "I record the sounds and play them back, and the first reaction of the people is, usually: wow, I didn't know that sound did that."[18] Instead of exclaiming "I didn't know that *I* did that," Oliveros reports that her students immediately granted an agency to sound that emerged only in the act of listening to their improvisation with tape. The feedback from this system helped students gain a knowledge of their body's relationship to their instrument and to sound with each iteration, but the goal for Oliveros was not technical mastery—it was instead a mode of interaction with the instrument that would allow new sounds to emerge without a hierarchical model of control.

Oliveros brought this practice of improvisation with tape into her life as a public performer, as well. In 1960 she began to improvise with Ramon Sender, another student of Erickson's, who had been building an electronic music studio in the attic of the SFCM. Improvising in this studio proved different from improvising at KPFA, since rather than having just one master tape machine, Sender's studio also had rudimentary equipment that he hoped could approximate the "classical" electronic music studios he had read about in Cologne and Paris.[19] It was in this studio, and at her home—with her one-track tape recorder—that Oliveros began to expand her experimental systems to include both acoustic and electronic instruments. It was also at the SFCM where she began to expand the practice of improvisation with tape from a purely pedagogical practice to one that could be performed in front of an audience.

Oliveros's shift from improvisation as compositional technique to improvisation as performance proved to be a minor scandal in the relatively conservative institutional music milieu of San Francisco. In 1961, Oliveros and Sender initiated the *SONICS* concert series at the SFCM, including not only performances of compositions but also live performances of improvisation with tape, spatialized throughout the concert hall and accompanied by light projections that reacted to the music, emphasizing their real-time spontaneity.[20] Although the composer Fred Frith has characterized Oliveros's early improvisations as "not concerned with [their] place in history," Oliveros must have been aware of the significance of presenting improvisation at the SFCM, especially on December 16, 1961, the 191st birthday of Ludwig van Beethoven.[21] A television news segment about the concert aired on local channel KPIX intercut footage of Oliveros, Sender, and Johnson improvising with footage of a separate performance of a Beethoven string quartet, ending with a menacing zoom onto a bust of Beethoven; a preview in the *San Francisco Examiner* asked, "Is this really the way to celebrate Beethoven's birthday?"[22]

Oliveros's performance of live improvisation with tape was not simply a public recreation of her private small group practice. Instead of the group all recording to a single tape machine, each musician recorded their signal to their own separate machine, and the signals from these tapes were routed to a spatialized loudspeaker system placed throughout the hall; the signal of each individual musician would be fed back through the hall to various members of the group, effecting a real-time cybernetic circuit.[23]

According to the concert's press release, "a feature of the program will be live improvisation by the four composers, each recording on a separate machine. Regulated by a keyboard that plays 'spacial [*sic*] relationships,' the sounds will be played back simultaneously through 15 speakers placed strategically throughout the auditorium, in the corridors leading to the auditorium, and in the adjacent courtyard."[24]

Milton Cohen, the designer of this spatialized speaker system, would later write that he wanted his installation to "suggest a museum of creative presence, of living performance, of spontaneous action."[25] That such spontaneous action would only be possible through the systematic and cybernetic routing of sonic signal from human musicians to tape machines to loudspeakers and back to those humans suggests that Cohen, like Oliveros, was also deeply invested in the promise of cybernetics to valorize "living performance" in real time.

By expanding the patchwork of her instrumental system to include the human body as one object among others, Oliveros engendered a new kind of subject out of a system of organs: not only the flesh of her mouth and hands on the metal of the horn, but soon, the very electric potentials of her brain's synapses onto the electric waveforms generated by new electronic instruments, connected by patchwork. Free improvisation with tape suggested to Oliveros that her musical self comprised different consciousnesses in her mind and body, revealing the sound of gestures rendered audible through their incorporation into the system. Quite literally incorporating new electronic instruments into a patchwork musical self in Sender's attic studio, her own home, and very soon thereafter at the newly formed SFTMC, Oliveros would further develop new experimental systems with new protocols, more explicitly translating technoscientific concepts into experimental practice. This practice of self-discovery produced audible artifacts that show its development throughout the 1960s.

The Body Electric

Although the majority of Oliveros's tape pieces from the 1960s were produced between 1965 and 1967, she also produced two early works of electronic music that did not use the instruments of a typical electronic music studio. Instead, these pieces used the human body itself as an "electronic" instrument. From 1959 to 1961, Oliveros worked on a fixed tape piece entitled *Time Perspectives*, both in the SFCM's attic studio and in her own home. To produce sounds for this work, Oliveros recorded the sounds of her body playing domestic objects as percussion instruments and used other domestic objects to modulate those sounds: cardboard tubes for filters, a bathtub for a reverb chamber, and her own technique of variable speed recording and playback, which she effected by controlling the tension of tape reels with her hands.[26] In *Time Perspectives*, the human body is expressly part of the instrumental system that produces electronic music, rendered as music that explores the potentials of the patchwork body within a system of instruments, be they cardboard tubes and bathtubs or metal boxes filled with electronic circuitry.

Very soon after this first tape was completed, Oliveros translated the sounds of this new kind of musicality to a notated piece for mixed a cappella chorus entitled *Sound Patterns*. In this work, as in her very first work of electronic music, human bodies were posited as part of a system of electronic signals traveling from brains to lungs and larynxes, producing sounds that approximated the electronic effects of white noise, ring modulation, voltage envelopes, and frequency filters, as if those electronic instruments had always already been parts of a human body.[27] *Sound Patterns* was awarded the Gaudeamus International Composers Award in 1962, and in September of that year Oliveros traveled to Bilthoven, in the Netherlands, to receive it.[28] In her absence, Sender and Subotnick's new SFTMC was quickly acquiring equipment, and notoriety, in the Bay Area. When Oliveros returned in 1963 to their new location at 321 Divisadero Street, she was greeted with multiple tape machines and many new instruments. Like Sender's attic studio at the SFCM, the SFTMC at 321 Divisadero Street contained an odd assortment of equipment and was in a constant state of flux throughout 1964. In this ever-changing studio, Oliveros began to develop new performance practices that would more concretely incorporate her body into patchwork systems with flexible boundaries.

PATCH 2: HUMAN/ACCORDION/MYNAH BIRD/SEESAW

After Oliveros's return to San Francisco in early 1963, several events would lead her to her "experiments of a patchwork quilt nature" with oscillators in early 1965; one of the most significant was her development of a relationship with the pianist David Tudor. After meeting at a dinner party hosted by Olive Cowell, the composer Henry Cowell's aunt, in the summer of 1963, Oliveros and Tudor began a long friendship and correspondence, bonding over their mutual affection for free reed aerophone instruments such as accordion and bandoneon.[29] In the fall of 1963 Oliveros began planning a festival known as the Tudorfest at the SFTMC—a multiday festival featuring David Tudor performing the works of the New York School, including pieces by John Cage, Toshi Ichiyanagi, George Brecht, and Tudor himself. Many of these works involved the use of electronics for live performance, and Oliveros's correspondence with Tudor throughout late 1963 and early 1964 evinces her deep involvement in the festival's technological planning.[30] Oliveros also began to compose her own piece for the festival, provisionally entitled *Duo for Accordion and Bandoneon*. This was to be a purely acoustic composition with a traditional score and no electronic elements, but through the process of rehearsal and performance, it turned into an improvisation that highlighted Oliveros's embodied cybernetic relationship with her instrument.

To compose her *Duo*, Oliveros used her practice of improvisation with tape to "discover her natural compositional inclinations," writing out an iterative score using a combination of fixed and indeterminate notational systems. Listening in on these sessions in Oliveros's apartment was Ahmed, her partner Laurel Johnson's pet mynah bird, who had learned to mimic and vocalize the sound of Oliveros's accordion and tape machine (see figure 10). As Oliveros wrote to Tudor of these rehearsals, "Ahmed has made a definite bid to be a member of this performance. Every time I pick up the squeeze box or play the tape he joins very positively (I thought it would be appropriate to have a CAGE on stage even if it wasn't John). Also he is producing fantastic combination tones now!"[31] In addition to the feedback of the tape, Oliveros found herself with an avian source of feedback as well. Oliveros's anecdote offers a poetic and prescient recollection of the function of the mynah bird in Aldous Huxley's popular 1962 novel *Island*, in which mynah birds appear throughout the utopian island repetitively screeching "here and now!"[32] Ahmed's synchronous feedback indeed functioned as a reminder for Oliveros to be "here and now," radically increasing the pace of her extant practice of self-study with tape, forcing her to be

Figure 10. Pauline Oliveros, David Tudor, and Ahmed the mynah bird on March 25, 1964. From Fang family, San Francisco Examiner photograph archive © The Regents of the University of California, The Bancroft Library, University of California, Berkeley. This work is made available under a Creative Commons Attribution 4.0 license.

acutely aware, in the moment of performance, of the sounds that emerged from her body and her instrument.

In addition to the sonic feedback from Ahmed, Oliveros also asked the dancer Elizabeth Harris, with whom she had worked in the early 1960s, to create some kind of mechanism for creating physical feedback during the performance as well.[33] When Oliveros and Tudor arrived at the SFTMC for dress rehearsals, they discovered that Harris had constructed a large, homemade seesaw that moved vertically and rotationally and had swiveling seats with seat belts, on which Oliveros and Tudor would balance during the performance.

Not only did this apparatus make it impossible to read from the score Oliveros had prepared; it also fundamentally changed Oliveros and Tudor's relationship to their instruments. Pulling and pushing their instruments' bellows would propel them throughout the space, sometimes pulling Oliveros off the seesaw because of the accordion's asymmetrical weight.[34] Buckled in to the seesaw apparatus, her hands strapped to the accordion (which was in turn affixed to her body), Oliveros was forced to constantly take stock of her physical body as it moved in space, cybernetically adjusting its performance in an attempt to both perform her written score and not fall off the seesaw. What began as a semi-improvised, semi-indeterminate composition became an experience in which Oliveros had to cede a certain amount of control over her instrument and over her own body's movement in space, and also to Ahmed, an unpredictable feedback source patched into this system. The final program recognized this agential extension by listing the piece as *Duo for Accordion and Bandoneon with Possible Mynah Bird Obbligato, See-Saw Version*.[35]

Although Oliveros's *Duo* did not use electronics, the rest of the Tudorfest was saturated with them. Photographs from the event show tables full of amplifiers, tape recorders, and other unidentifiable black boxes, possibly for use in works by Cage and Ichiyanagi.[36] On the third night of the festival, Oliveros helped perform Cage's *Atlas Eclipticalis* with *Winter Music, Electronic Version* and *Cartridge Music*, both of which involved the use of live electronics. This experience with electrical instruments (contact microphones, amplifiers), as well as her experience with tape, began to suggest an electronic metaphor to Oliveros: her physical body could be imagined as only one informational node in a circuit of instruments, oscillating up and down on the seesaw, reacting to emergent sonic phenomena in unpredictable and increasingly uncontrollable ways, producing sound that could be inscribed as magnetized iron coated on acetate, transparently reproduced by loudspeakers. At the Tudorfest, sound circulated as a physical waveform in the air, an electrical signal on tape, and a perceptual signal in human consciousness. The electrical and electronic instruments used by Tudor at the Tudorfest, and newly available to Oliveros at the SFTMC, promised to extend, increase, and complicate this control. After the Tudorfest, Oliveros quickly began her own experiments to explore this new kind of musical performance.

"Research of a Patchwork Quilt Nature"

The SFTMC's studio changed dramatically after the Tudorfest. As of the time of the festival, as described in chapter 2, its equipment mostly consisted of mismatched military and telephone surplus equipment, with a variety of half-finished and stalled projects that failed to solve its major technical challenges. During the summer of 1964, however, the center's second technician, William Maginnis, rearranged the studio to make it more ergonomic and amenable to connections between devices; instruments were placed on waist-high equipment racks, now within arm's reach, easily patchable to each other. In addition, Maginnis solved many of the studio's electrical issues, such as mismatching impedances and ground loops, possibly with help from Don Buchla, from late 1964 into 1965. At some point in 1964, the SFTMC had also acquired a smaller Ampex PD-10 tape duplication system, which comprised four separate tape transports on a single rack with exposed and easily modifiable erase, record, and play heads (see figure 11). For the first time, instruments in a single rack could be easily patched together to form ad hoc systems for the cybernetic routing of electronic and sonic signal.

With this new studio, Oliveros began a new project of developing live electronic music with her own body as an electrical component in the patchwork system, although she lacked the language to describe what, exactly, she was doing. Throughout late 1964 and early 1965, she was still composing notated works, notably the theater piece *Pieces of Eight*, which involved a large papier-mâché bust of Beethoven with glowing red eyes; but she was also beginning to experiment with oscillators and tape systems recently installed in the SFTMC. It was during this period, in early 1965, that she began her "research of a patchwork quilt nature" with the studio's oscillators, tape machines, and patch bay.

This early research produced patches between body and electronic instruments that Oliveros later fabricated into a cohesive narrative about the long arc of her career. Indeed, later in her life, the radical contingency of

Figure 11. Photograph of equipment at the SFTMC, ca. late 1964, by William Maginnis, shown to author on August 23, 2016. On the center rack, two matching Hewlett-Packard 200CD oscillators, another unidentified oscillator, and an oscilloscope are visible. The right rack is entirely occupied by the Ampex PD-10 Tape Duplication System, with four tape transports in a vertical configuration.

this moment became quilted into her own self-narration, an experience like the one she had with her tape recorder in 1953, which she characterized as a breakthrough in improvisation:

> A big breakthrough came when I discovered how to improvise with classical electronic studio equipment in 1965. The system I used to create electronic

music consisted of tube signal generators, a patch bay, line amplifiers, and two stereo tape machines. [...] The huge dials that had seemed so unfriendly to performance now became receivers for the musical knowledge embodied in my hands and fingers. I had created a very unstable nonlinear music-making system. [...] I had created a new musical instrument that included my own bodily input in the form of gestures mapped onto the machine for analog output to speakers. This meant that I could play my electronic music in real time without editing or overdubbing.[37]

Oliveros remembers this experience using the language of cybernetics and systems theory, using terms such as *instability, nonlinearity, systematicity, input,* and *output* as if they were native to the domain of music. Mapping bodily gestures to sound, this "very unstable nonlinear music-making system" had some similarities to Morton Subotnick's fantasy of a composer's black box; it valorized only its input and output, making it unnecessary to understand the complexity and nonlinearity of its inner workings. However, Oliveros's system featured one crucial difference. While Subotnick's speculative black box sought out more complete control of electronic sound by its human user, Oliveros's black box distributed the control of its output throughout its various components, both fleshly and machinic. Oliveros understood her system, which was nearly impossible for a human to control with any sense of precision, accuracy, or even predictability, as a way to research her body's consciousness, a deeper sense of self that could not be intentionally controlled. Although this body consciousness was still her own—and thus not as opaque and unknowable as the agency that Sun Ra would claim controlled his performances with his black boxes, as discussed in chapter 5—it was still far from the "heroic vision" that Subotnick wanted to extend through his black boxes. In this sense, the "very unstable nonlinear music making system" described later acts as a hinge in the narrative arc of this book: a cybernetic system wherein the human is neither completely in control nor completely black boxed.

PATCH 3: HUMAN/OSCILLATORS/PATCH BAY/ TAPE MACHINES/AMPLIFIERS/SPEAKERS

At this point we return to the vignette with which this chapter opened: Pauline Oliveros staring at oscillators in the SFTMC's attic studio in 1965.

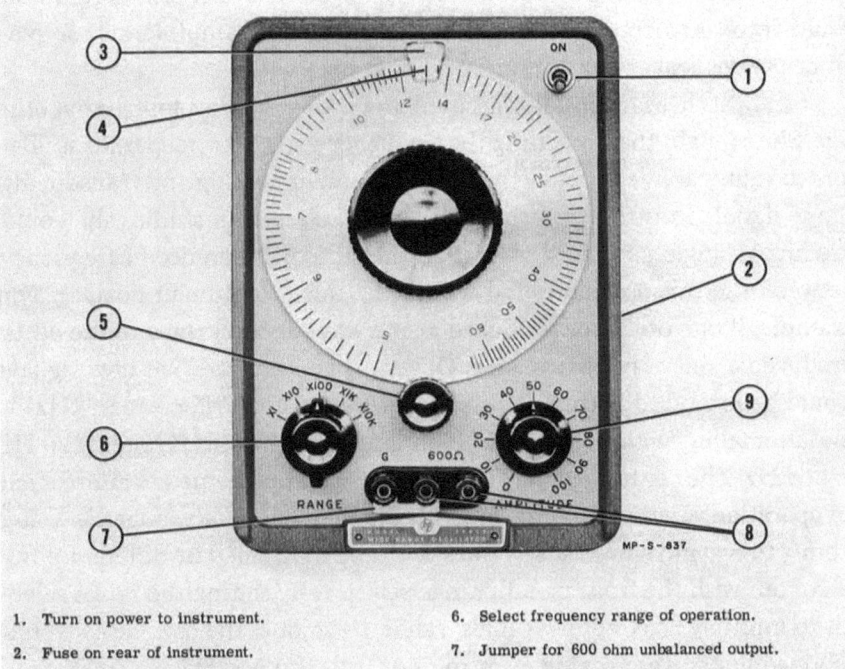

Figure 12. Illustration of the HP 200CD wide range oscillator, with instructions for use, from its 1955 manual.

The previous year, she had seen David Tudor use homemade electronic black boxes to perform live electronic music, without tape overdubbing or splicing, at the SFTMC's Tudorfest and had even made a physical black box for her 1964 composition *Apple Box*, wherein she affixed contact microphones to a wooden apple box and activated the box by striking and rubbing it with various materials, electrically amplifying acoustic sound without translating it into informational electronic signal.[38] But sitting in front of the SFTMC's oscillators, Oliveros was thinking about something very different than amplifying acoustic sounds. She wanted to explore the particular affordances of the "huge dials" on the front panels of the HP 200CD oscillators (see figure 12) in order to play with a signal that was at once acoustic, electronic, and perceptual, a unified energetic signal that could travel in a feedback loop between oscillators, amplifiers, tape machines, loudspeakers, ears, brain, and fingers.

Staring at these "intimidating" oscillators, Oliveros developed a hypothesis. She thought that by setting the frequency dials of two separate oscillators to values above or below the range of human hearing and transducing those signals into sound through loudspeakers in the studio, she could produce a single audible signal when the sum or difference in frequency between the two oscillators fell within the range of human hearing. For example, if one oscillator produced a sine wave at 100 kHz, and the other produced a sine wave at 105 kHz, Oliveros hypothesized that new signals would be produced both at their sum (205 kHz) and difference (5 kHz)—the latter falling within the range of human hearing, which is roughly 20 Hz to 20 kHz. The controls of the HP 200CD oscillators seemed to afford such an operation, since their large, logarithmic frequency dials ranged in value from 5 to 65, and since those values could be multiplied by different powers of ten with the front panel's range selector. By setting the range selector to multiply the frequency dials' values by 10,000, the oscillators would produce sine waves ranging from 50,000 Hz to 650,000 Hz; since the dial's scales were logarithmic, a small turn at their lower range would change their frequencies by a small amount, but that same amount of physical travel would produce a much larger change at their higher range. This meant that a single sweep of the dial on one or both oscillators would theoretically produce difference tones that quickly passed from infrasonic through sonic to ultrasonic ranges at differential rates depending on the position of the dials, putting an unpredictably wide range of frequencies at Oliveros's trembling fingertips.

Oliveros's hypothesis relied on a tacit translation between a number of related but distinct phenomena in the domains of human auditory perception, acoustic sound, and electricity: the psychoacoustic phenomena of combination and difference tones, the acoustic phenomenon of acoustic beats, and the electrical phenomenon of intermodulation. As an accordionist, Oliveros was very familiar with combination and difference tones; when two adjacent notes are played on an accordion at loud volumes, a new tone is often perceived with a fundamental frequency equal to the sum and/or difference of the fundamental frequencies of the first two notes.

These combination and difference tones, sometimes called "Tartini tones" after the violinist who popularized them in the eighteenth century, are psychoacoustic; they are perceived by the human auditory apparatus but are not produced acoustically as sound waves in air. The accordion can, however, easily create the related but distinct phenomenon of acoustic beats, or "beat frequencies"; these are readily produced by depressing two adjacent keys, especially in the instrument's low register. The low frequency acoustic interference in air produces a fluctuation in the amplitude of the accordion's overall sound production, perceived as a "beat" with a frequency equal to the difference between the fundamental frequencies of the two keys.[39]

Electrical intermodulation, also known as intermodulation distortion, is fundamentally different from psychoacoustic combination/difference tones and acoustic beats, although it is described with similar mathematical expressions. It occurs when two or more signals are combined in a nonlinear system, producing a number of additional signals. Depending on the particular nonlinearities of the system, these additional signals have frequency components at the sum and difference of the two original signals, as well as at the sums and differences of multiples of the original signals. Expressed mathematically, when two sine waves at f_1 and f_2 are subjected to intermodulation distortion, the system will produce signals at $f_1 + f_2$ and $f_1 - f_2$ (their sum and difference) and also at $2f_1$ and $2f_2$ (their harmonics at the octave). These phenomena are known as second-order intermodulation products. However, intermodulation distortion also occurs at a third order when these second-order products intermodulate with the original signals themselves, producing signals at $2f_1 + f_2$, $2f_1 - f_2$, as well as $f_1 + 2f_2$ and $f_1 - 2f_2$, the latter of which often come very close in absolute value to the original two frequencies.

Some electrical devices, such as ring modulators, are explicitly designed to intermodulate the frequencies of two or more signals and produce linear outputs, such as at the sum or difference of the two original signals, while rejecting the input signals themselves along with their second- and third-order intermodulation products. But intermodulation distortion regularly occurs in many electrical devices that are not designed for it, due to the physical nature of their component materials, especially when signals are highly amplified. (This is known as the "rusty bolt effect," when some oxidized or corroded part of a system acts as an unintentional source of passive intermodulation distortion.) Although the SFTMC did house a ring modulator, Oliveros did not use it in her system; she seemed to be unaware of its necessity for electrically intermodulating two oscillators to produce what she called "combination and difference tones." Instead, Oliveros assumed that the psychoacoustic and acoustic phenomena of difference tones and beat frequencies could be simply replicated using electronic oscillators alone. This relied on an understanding of sound as an informational signal that was fungible across psychoacoustic, acoustic, and electronic domains. In Oliveros's imagined system, sound, as an informational signal, would complete a recursive circuit between mind, body, oscillator, loudspeaker, air, and sonic perception. Since Oliveros was both unable to intentionally control the system with the unwieldy oscillator dials and unable to account for the system's behavior when the dials were turned, the system ostensibly acted as a cybernetic circuit in which sound could act for itself—just as it had acted for itself in her practice of improvising with tape. This time, however, Oliveros would not need to stop and rewind the tape to the system's feedback; instead, it would occur in real time.

The emergent nonlinear behavior of Oliveros's system became immediately apparent when she first attempted to create psychoacoustic combination and difference tones by patching her two HP 200CD wide-range oscillators into amplifiers and then into loudspeakers. With the system patched and powered on, Oliveros set each oscillator's dial into the ultrasonic range and swept both dials through the wide frequency spectrum, thinking that they would produce difference tones. But during this first experimental attempt, no audible sound was produced. This could have been for a number of reasons; the most likely is that the loudspeakers in the SFTMC's studios were not able to produce sound waves above 20 kHz, the upper limit of human hearing. Since no acoustic sound was produced in the room to begin with, no psychoacoustic difference tones could be perceived.

Thinking that the tones were just very quiet, Oliveros patched each oscillator directly into an amplifier and then electrically summed the two amplified signals in the studio's patch bay—something the passive patch bay was not designed to handle. This electrically summed and amplified signal was then sent through an additional stage of amplification before being transduced into air pressure waves by the studio's loudspeakers.[40] Doubly amplifying these two signals and electrically summing their signal in a passive patch bay meant that Oliveros was massively overloading the electrical components in this circuit—vacuum tubes and transistors—which luckily did not melt, explode, or otherwise fail.[41] Instead, the nonlinear behavior of these components, given their extreme operating conditions, made them act as an ad hoc electrical mixer via the "rusty bolt effect," or in this case the "World War II–era military surplus patch bay effect," which produced a variety of additional signals that Oliveros did not anticipate. In addition to second- and third-order intermodulation distortion products, Oliveros's system also produced harmonic distortion, creating complex signals with harmonics at even and odd multiples of their fundamental frequencies. And in another emergent twist, when printed onto magnetic audiotape, the oscillators' supersonic signals physically intermodulated with the tape's preprinted supersonic bias signal, introducing another layer of intermodulation that did not correlate in a linear way to Oliveros's hands on the oscillator dials.

With this precarious, electrically dangerous system, the slightest touch of either of Oliveros's hands, intentional or not, would produce massive, reverberant, sweeping sounds. As she recalls, "Then I started touching the dials and saw that I could sweep the audio range just by moving my hand a little bit and pretty soon I had a musical instrument that I could play. I could make sounds that didn't take much movement but it took a lot of listening to understand how to play this new instrument."[42]

Oliveros's system was incredibly difficult to control. Unlike an acoustic instrument, the sonic output produced by her gestural input changed every time she used it. In order to understand how to play this new instrument, Oliveros turned to the same disciplined protocol that she had been developing since 1959 of improvising, recording the improvisation to tape, rewinding the tape, listening back, and discovering what happened. But unlike the simple one-track tape recorder she had been using for the previous five years, she now had access to the SFTMC's newly procured Ampex PD-10 tape duplication system, located in the studio immediately to the right of the two HP 200CD oscillators, which dramatically expanded the role of magnetic tape in the development of her new practice.

This Ampex PD-10 tape duplication system, which had fallen off a shipping pallet and was too damaged to sell, was unique: it comprised four (not-so-)synchronized tape transports arranged vertically in a single rack, each with exposed record and playback heads that could be used in monaural, stereo, or quarter-track configurations, independently operating at speeds of 15 ips down to 1 7/8 ips. This meant that Oliveros could easily print the signal of her oscillators onto tape with one machine and then send that signal to one or more other record and playback heads on other machines, creating a cascading system of delayed signal. When all of these various delayed signals were monitored on the studio's loudspeakers, Oliveros would hear the oscillators's intermodulated outputs both in real time and also with short time delays, equal to the amount of time it took the printed signal to travel between tape heads. If a tape with her signal was sent back again to a playback head through the use of a tape loop, or by routing the tape to another nearby PD-10 transport, the "output" of the system would become its "input," feeding back signal both into the precariously saturated magnetic tape and through the loudspeakers into her own ears.

Without the incorporation of this tape delay system, Oliveros's pair of oscillators would only produce a signal that would be perceived as a single tone when transduced to air, something like a sweeping, distorted glissando with multiple harmonics of its fundamental frequency. Small, slow movements of her hands on the dials would not be easily perceptible because the signal was continuous, ceasing to be heard only when it fell below or above the lower or upper limits of human hearing. But with the incorporation of tape delay, any change in the signal from Oliveros's oscillators was audibly repeated at intervals of ~250 milliseconds and again at longer intervals, depending on how she routed the tape through the PD-10 system; the smallest motions of her hands on the HP 200CD frequency dials would produce massively reverberant, noisy, unpredictable behavior that became audible through its systematic delay and reiteration through Oliveros's self-regulating, cybernetic system.[43]

Mnemonics

Oliveros's development of a patchwork self can be readily heard in the first tapes that she produced with her experimental instrumental system of two HP 200CD oscillators, each with its own amplifier, patched and summed into the SFTMC's patch bay, with their combined signal routed to the record head of a stereo tape machine, likely a transport of the PD-10, further intermodulating with the tape's bias frequency. The recorded signal, after traveling the few centimeters from the record to the playback head, was routed back to the record head of a second tape machine, sometimes on a separate stereo track, resulting in a "ping pong" delay that bounced between the left and right stereo channels of the tape. Between April 1965 and January 1966, Oliveros produced no fewer than eight tapes with this system in various configurations, described in a letter to David Tudor as "*Mnemonics* [Supersonics]."[44] Five tapes with the *Mnemonics* title survive in Oliveros's archive, numbered I through V, as well as two unlabeled tapes that use the same instrumental system; these seven surviving tapes range in duration from eight to twenty-four minutes, suggesting that their duration was determined by the capacities of 5-inch and 7-inch reels of audiotape recording at the standard speed of 15 ips.[45]

There are no scores for these tapes, nor any overarching structural form. Instead, they trace Oliveros's improvisatory encounters with an instrumental system that itself functions as a *mnemonic*, a device to aid her memory. We can hear Oliveros's experiments in observing her own body's signal in real time; she lets the oscillators drone, lets feedback build up, then rapidly twists the dials; she rapidly turns the range switch of the HP 200CDs oscillators, briefly introducing audible sine waves in addition to the intermodulated signals of their sum and difference; and she explores the ghostly amplitude-modulated sounds that occur at octaves of the oscillators' fundamental tones, especially in the higher ranges of their operation. This system produced much more than what Oliveros thought it would, allowing multiple unforeseen sounds to emerge. If this system was a mnemonic, it was one that did more than aid Oliveros's memory; it allowed her to perceive what was otherwise impossible to consciously observe and to produce what was otherwise impossible to intentionally produce, in the moment of performance through the cybernetic routing of perceptual, acoustic, and electronic sound.

To closely examine the emergent, unintentional sounds contained in the *Mnemonics* tapes and to introduce another way to observe them, here I briefly turn to spectrographic analysis, which visually represents the information contained in an audio recording. In 2017 I attempted to recreate

Oliveros's initial hypothesis about producing combination and difference tones with two supersonic oscillators in the electronic music studio of Stony Brook University. If two supersonic sinusoidal waves were ideally mixed, they would produce a single "difference tone" with a frequency value at their difference. The resulting signal would have minimal harmonic overtones, because sine waves lack harmonics; in theory, it would be legible on a spectrogram as a single, bright line, traveling up and down the frequency spectrum as the dials were turned. At Stony Brook, I patched two transistorized, stable oscillators into a Bode Ring Modulator—a device designed to intermodulate two signals and reject both the original signals and any second- or third-order intermodulation distortion effects—and recorded the resulting signal directly to a digital recorder instead of onto tape.

The signal in the spectrogram (see figure 13), represented by a dark line, descends diagonally from the upper limit of the audible spectrum to the lower limit, and back up again; this is the result of the absolute difference between the two oscillators approaching zero and increasing again after that zero crossing. The signal has only minor harmonic content, visible as fainter lines at rationally determined distances above the main signal, which has a much higher amplitude. Thus, with the proper equipment designed specifically to mix two electrical signals and only output their first-order intermodulation products, I was essentially able to demonstrate one element of Oliveros's hypothesis that difference tones could be produced electronically.

The rest of Oliveros's hypothesis, however—that such tones were fungible between psychological perception, acoustic sound, and electricity—was disproven both by the fact that the experiment did not initially work in the SFTMC studio and by the multiplicity of emergent, nonlinear signals that were printed onto tape when Oliveros doubly amplified and summed her oscillators in the SFTMC's patch bay. These signals are clearly visible on a spectrogram of the first *Mnemonics* tape (see figure 14).

In this spectrogram, the output of Oliveros's system is seen as a dark line with fainter lines above it, which represent both normal harmonic distortion and the second- and third-order intermodulation products that do not relate harmonically to the fundamental frequency of the signal. In theory, it also includes signals produced by the intermodulation of a third, fixed signal—the bias signal printed on magnetic tape, usually anywhere from

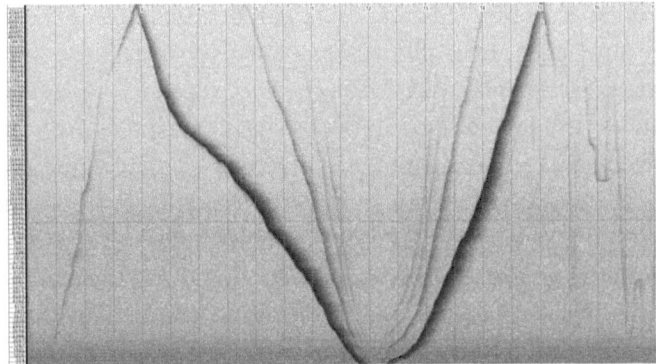

Figure 13. Spectrogram of Oliveros's oscillator hypothesis made by the author at the Stony Brook University Electronic and Computer Music Studio, September 7, 2017.

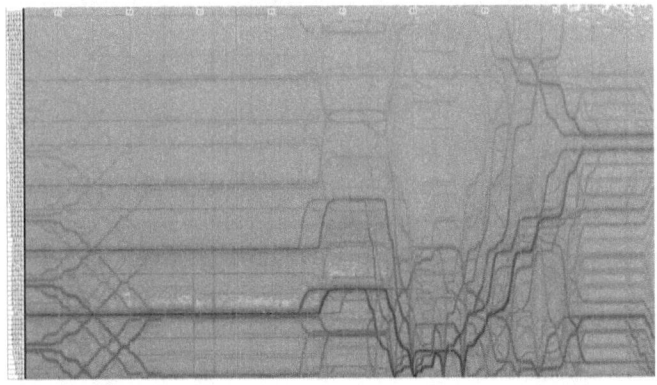

Figure 14. Spectrogram of nonlinear emergent signal dynamics in *Mnemonics I* (1965).

40 to 150 kHz. Since this signal had a fixed frequency, it may account for the "mirroring" effects that can be seen around certain frequencies in the spectrogram, making the output of this system sound like it was simultaneously descending and ascending in perceived pitch.[46] Oliveros's use of tape delay also meant that the effect of each turn of the oscillator's dials was repeated within a fixed time period, which can be observed in the ghost-like traces of signals from left to right in the spectrogram. Listening to the

Mnemonics tapes makes this readily apparent: the highly tape-saturated tone of Oliveros's system sounds distorted, noisy, with abundant tape hiss and precarious tape delay. Musically, these tapes sound alive, dangerous, unpredictable, and joyous, quite unlike a simple scientific experiment.

The *Mnemonics* tapes document Oliveros's "research of a patchwork quilt nature," an approach to music that sought out the emergent behaviors of patchwork bodies engaged in cybernetic, self-regulating performances. What Oliveros later characterized as "bodily input in the form of gestures mapped onto the machine for analog output to speakers" was not exactly legible during the experiment itself; it was very difficult for Oliveros to consciously control her own "bodily input" since the smallest involuntary twitch of her fingers would produce a cascade of signals traveling through an unpredictable, nonlinear system. This was the sonic output of a patchwork body, a uniquely emergent, temporary incorporation, contingent on the materialities of these particular oscillators, amplifiers, patch bay, and speakers at the SFTMC. Indeed, the sonic output of Oliveros's experiment changed dramatically when Oliveros later tried to execute this same experimental protocol with the Buchla Box, which was engineered to minimize the unintentional intermodulation and distortion of audio signal (even as similar modulation effects could be achieved if patched intentionally).[47] Like her earlier protocols of improvisation with acoustic instruments and tape for feedback, Oliveros's protocol for her electronic system valorized listening but also dramatically reduced the degree of human control to a few centimeters of rotation on the oscillator's dial. This meant that this system, perhaps even more than the Buchla Box, produced a self-regulating flattening of agency between human and instrument, a dance between two bodies, equally electric and equally complex.

PLAYING THE BODY ELECTRIC: SOUND AND CYBERNETIC AGENCY

Between 1965 and 1967, Oliveros produced dozens of tapes with different configurations of new instrumental systems—not only at the SFTMC, but also at the University of Toronto Electronic Music Studio (UTEMS) during a summer course in 1966 and as the director of the Mills College

Tape Music Center, which absorbed the SFTMC later that year.[48] In the summer of 1967, Oliveros held a "Tape-A-Thon" concert that presented thirty tapes at the quasi-communal warehouse space of the artist Ronald Chase, accompanied by visual displays by her then partner Lynn Lonidier and also by Chase's films starring her ex-partner Elizabeth Harris.[49] Over the course of twelve hours, audience members heard the audible traces of Oliveros's unpredictable, emergent performances within cybernetic systems in which sound seemed to act as a signal that regulated Oliveros's emergent patchwork bodies, each exhibiting a unique performance dynamic in relation to its specific technicity. It was one of the only public performances of the taped artifacts from Oliveros's early electronic music practices in Oliveros's lifetime; these tapes would not be heard again for decades.[50]

In 1967 Oliveros moved to San Diego at the invitation of Robert Erickson to join the faculty of the newly formed music department at the University of California San Diego. There, as Kerry O'Brien has explored, Oliveros's music saw a marked shift in its aesthetics. Instead of the noisy and unpredictable tapes produced with her experimental electronic music systems throughout the previous two years, Oliveros's work at UC San Diego became more contemplative and meditative, largely in response to the violence of the Vietnam War and the assassination of Robert F. Kennedy, as well as Oliveros's deepening interest in movement practices such as the Alexander Technique and martial arts.[51] This shift led Oliveros to create the widely celebrated *Sonic Meditations*, protocols for studying the body and its relation to sound, developed with a gynocentric community named the ♀ Ensemble beginning in 1970. Although Oliveros had never hidden her identity as a queer woman, she began to incorporate her queer identity into her music more visibly in this period of her career. Not only was the ♀ Ensemble "purposely all female in order to maintain a common, stable vibration within itself and to explore the potentials of concentrated female creative activity," but in the first publication of the *Sonic Meditations* in 1971, Oliveros described herself as a "two-legged human being, a female, lesbian, musician, composer, among other things which contribute to her identity."[52] This body of work, as Martha Mockus has argued, is explicitly feminist, calling for women to queer the perceptual processes of their bodies and consciousnesses in new ways, producing what Mockus calls a "lesbian musicality" in which music is "not an object but a process engaging bodies, time, and space."[53]

Because Oliveros's long-standing exploration of gender, sexuality, and sonic embodiment began to flourish in the 1970s, journalistic and scholarly appraisals of her earlier electronic music have often focused on a single tape bearing the title *Bye Bye Butterfly*, in which a recording of the aria "Un bel dì, vedremo" from Giacomo Puccini's 1904 opera *Madama Butterfly* was incorporated into the same system of oscillators and tape delay with which Oliveros produced the *Mnemonics* tapes in 1965. In Oliveros's memory, this tape was created when she simply grabbed a nearby record at random in the studio and incorporated it into her patchwork system to see what would happen; the incorporation of an aria from *Madama Butterfly* was purely coincidental. Although Oliveros produced this tape in 1965, it was evidently not important enough for her to include in a list of her extant tape works sent to David Tudor in early 1966; while it was played during the "Tape-A-Thon" in 1967 along with twenty-nine other early tapes, it remained otherwise unheard until 1977, when it was commercially released on an LP record of electronic music entirely created by women.[54] And although Oliveros apparently only submitted *Bye Bye Butterfly* to be included on this record because it was short (8 minutes), scholars such as Mockus "became convinced that *Bye Bye Butterfly*'s presence on this collection of all-women's music invites a feminist reading of the piece."[55]

The presence of an identifiably feminine voice in the patchwork system of this tape, although it was chosen accidentally, has readily allowed for hermeneutic readings that connect Oliveros's early electronic music with her later and more explicitly feminist work. The composer Heidi von Gunden, a student of Oliveros's who published the first monograph on her work in 1983, declared *Bye Bye Butterfly* "an easy piece to understand": "the tape-delay technique and frequency modulation produce wavelike gestures resembling sonic good-byes to [*Madama*] *Butterfly*."[56] Echoing this reading, Mockus has more recently written that "*Bye Bye Butterfly* enacts an eerie and forceful feminist critique of the opera," hearing the oscillators as "a sonic alarm alerting me of Oliveros's bold confrontation with Puccini's opera" and the tape delay system as the "rhythms of the sky, oceans, and wind" as well as the "pulse and vibrations of blood," evoking the character Butterfly's suicide. For Mockus, *Bye Bye Butterfly* "works to squash heterosexual-patriarchal stereotypes of women" through a reclaiming of the butterfly as a poetic and morphological representation of the human vulva—a "beautiful symbol of lesbian sexuality."[57]

There is no doubt that Oliveros's identity as a queer woman in the 1960s played a major part in her musical life. But to single out *Bye Bye Butterfly* because of the accidental presence of a feminine voice, and to listen to these tapes hermeneutically, as if Oliveros intentionally created them as linear compositions, perhaps misses a more radically feminist proposition of her early electronic work: that the human body is a patchwork body that can be changed through connection with new technological instruments, and that such connections produce emergent, self-governing performance dynamics that are distributed throughout the system. Oliveros's embrace of cybernetics was an embrace of the systematicity of the human in relation to sound, which she understood to be a fungible signal that connected human perception, the physical world, and the domain of electricity.

For Oliveros, the harnessing of this signal was a feminist act. In a 1968 article about her practices creating electronic music throughout the previous three years, Oliveros characterized her experiments with oscillators and tape delay systems as acts of witchcraft that tapped into the magical power of pure sound:

> When I was thirty-two, I began to set signal generators beyond the range of hearing and to make electronic music from amplified combination tones. I felt like a witch capturing sounds from a nether realm. In one electronic music studio I was accused of black art, and the director disconnected line amplifiers to discourage my practices, declaring that signal generators are of no use above or below the audio range because you can't hear them. [. . .] I worked there for two months, and, for recreation, would ride my bicycle to the town power plant where I would listen for hours to the source of my newly-found powers.[58]

Although Oliveros acknowledged the centricity of line amplifiers to her work, she still characterized her systems as outputting "amplified combination tones," positing a fungibility between psychoacoustic, sonic, and electrical phenomena. It was precisely the fungibility of sound as a signal that could travel between all three domains that formed the conceptual basis for Oliveros's early electronic work and continued to inform her work throughout the rest of her career. In these systems, sound seemed to act for itself, a powerful and dangerous signal that challenged unilateral compositional authority and promoted an exploration of the human body and its sensory, physical, and pleasurable relationship to sound.

Not only did sound seem to act for itself; for Oliveros, it became the basis for both human and nonhuman life. In the last sentences of that same 1968 article, Oliveros concludes: "The birds and insects share the air with waxing, waning plane and car drones. The insects are singing in the supersonic range. I hear their combination tones while the insects probably hear the radio frequency sounds created by motor drones, but not the fundamentals. If we could hear the micro-world, we would probably hear the brain functioning."[59]

Imagining the presence of sound in multiple kinds of life at multiple scales, Oliveros's early electronic work laid the groundwork for theorizations of sound that she would develop for the rest of her career. As Douglas Kahn has explored, Oliveros's eventual theorization of what she came to call the "sonosphere" "embraces a full sweep and barrage of energies, including the magnetic, electrical, electromagnetic, geomagnetic, and quantum, as well as the acoustical. It resonates among personal and interpersonal, musical, earth, and cosmological scales informed by physics and metaphysics."[60] The sound that permeated such a sonosphere was both informational and magical. In the 1974 publication of a more complete collection of her *Sonic Meditations*, for example, Oliveros posited a "Sonic Energy" that was transmissible both through individuals and within groups of people, allowing for both "communication among all forms of life" and "humanitarian purposes; especially healing."[61] In the 1970s, Oliveros's explorations of such sonic energy were often limited in scale to small groups of humans, or smaller—the individual human, or even their individual brain. Indeed, after experiencing an early 1970s performance of Alvin Lucier's *Music for Solo Performer*, discussed in chapter 4, Oliveros apparently commented: "We no longer needed electronic music studios; we already had them in our brains."[62] But in 1999, Oliveros's explorations of sonic energy grew in scale to the cosmic: "What would a spatial melody sound like—a pitch beginning on Saturn moving to Aldeberon to Sirius to Earth? Space related frequency and amplitude—multidimensional melody—color/space/sound melody. Who would be playing this tune? Who would be listening and where? Melody across space stretched out and also happening everywhere simultaneously. Space is the place—I hear you Sun Ra!"[63]

As chapters 4 and 5 explore, the conceptualization of sound as a fungible informational signal that allowed for the emergence of apparently self-governing performances proved to be an extremely powerful idea for both Alvin Lucier and Sun Ra, respectively. Like Oliveros, they developed new technologies and techniques for human-instrument interactions that challenged inherited notions of human subjectivity and human agency, especially of humans as bounded subjects who could communicate their innermost subjectivities through music. For Lucier and Ra, as for Oliveros, the human subject was instead conceptualized as a system with varying degrees of opacity, legibility, and observability—each of which carried differing social, political, and historical consequences for their musical practices in mid-century America.

4 Alvin Lucier and the Ambiguity of Sound and Signal

> Real echoes
> Direct imitation, without metaphor,
>   Without metaphor, but with ambiguity.
>
> —Alvin Lucier, sketch for *Vespers* (1967[?],
>   ALP, box 12, folder 4)

EEEK/LUNK/EEOWW/TUMP/RRROOOMM/OOOOP/
NOOZLE/LINGG/CLACK/MOOOOO/UGH/SHOOSH/
IZZLE/CLUMP

In the May 1967 issue of the *Sylvania Scanner*—the in-house magazine for Sylvania Electronics Systems, a subsidiary of General Telephone & Electrics Corporation and one of the largest military contractors in the booming Cold War defense industry—an article by Major General Harry James Sands, Jr. details his experiences negotiating with North Koreans in Panmunjom. "There was apparently no way of penetrating the Communists' party-line scripts or reading their pre-conditioned minds," he laments. These communist minds, to Sands, weren't playing fair. Not only did they refuse to engage in good faith negotiations, which could be studied and possibly predicted, but they simply refused to be understood. Sands concluded by asserting that "only one aspect of the free world can really be communicated to the Communist mind—strength."[1]

Sands's article treats the "Communist mind" as an impenetrable black box that can only understand the threats, maneuvers, and tactics of military strength. His complaint about the impenetrability of the communist mind implied that it wasn't acting logically or predictably, as he had been trained to expect a human enemy to behave. As the historian of science Peter Galison has argued, this understanding of a human enemy as an impenetrable black box who nevertheless acts in accordance with predictable behavioral logic emerged in the 1940s, when the emerging discourse of cybernetics was employed to design antiaircraft weapons that would predict enemy pilots' behavior. In what Galison terms the "ontology of the enemy," the enemy was imagined as "at home in the world of strategy, tactics, and maneuver, all the while thoroughly inaccessible to us, separated by a gulf of distance, speed, and metal[.] [This] was a vision in which the enemy pilot was so merged with machinery that (his) human-nonhuman status was blurred."[2]

Galison argues that this ontological vision of enemy behavior quickly and recursively fed back onto the understanding of Allied soldiers themselves, and from there, "it was a short step from this elision of the human and the nonhuman in the ally to a blurring of the human-machine boundary in general. The servomechanical enemy became, in the cybernetic vision of the 1940s, the prototype for human physiology and, ultimately, for all of human nature."[3] Thus, for General Sands, the only dimension of behavior that could be legible between minds in the "free world" and minds in communist North Korea was not metaphorical strength, but physical strength, measured in troop numbers, warhead payloads, and megatons.

But in this same issue of the *Scanner*, on the very next page, an article's headline presents something else emerging from the mind, almost resembling the sounds of munitions fired at the enemy: "Eeek / Lunk / Eeoww / Tump / Rrrooomm / Oooop / Noozle / Lingg / Clack / Mooooo / Ugh / Shoosh / Izzle / Clump."[4] This article posits that the human brain, in addition to being a calculating war machine, is also a musical instrument: "Stradivarius violins are now second fiddle. A recent discovery reveals 3 1/4 billion musical instruments more delicate and more finely tuned than the grandest of 17th century violin-maker's creations. The human brain (there are approximately 3 1/4 billion such devices now existent) is emerging as a 'prima voce' among the bloops and bleeps [...] of the rapidly expanding new world of electronic music."[5]

The article continues to describe the novel work of the composer Alvin Lucier, who had recently begun experimenting with electroencephalogram (EEG) equipment given to him by Air Force researcher Edmond Dewan,

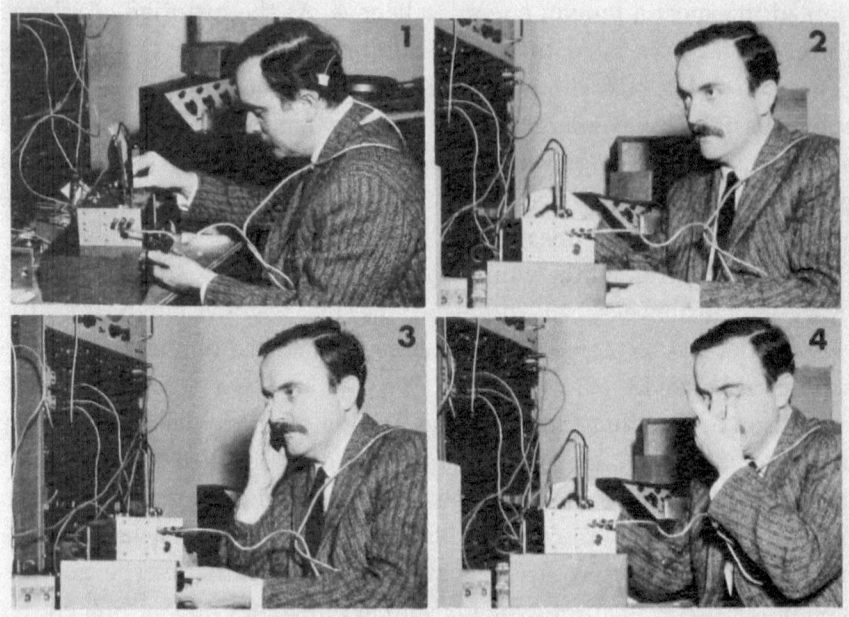

Alvin Luciers brainwave emanations: (1) Equipment in place and adjusted; (2) "Off"; (3) "Monaural"; and (4) "Full Stereo."

Figure 15. Photograph of Lucier's experimental protocol with EEG, in *Sylvania Scanner*, May 1967. From ALP, box 1, folder 1.

and who was soon to be commissioned by Sylvania to create a musical work with their vocoder (a portmanteau of "voice encoder"), a device developed for encrypted wartime communication. It describes Lucier as a gentleman scientist engaged in self-experimentation with a novel, electronic apparatus, framing his experiment in carceral terms, perhaps in an attempt to connect it to a topic more familiar to the *Scanner*'s readership: "In a microminiature studio/laboratory at Brandeis University in Waltham, Mass, composer Alvin Lucier, looking every bit like a condemned man seconds before the warden throws the switch, sits calmly, electrodes at his skull, recording his brain waves."[6]

In a series of four sequential photographs (figure 15), Lucier is shown executing his protocol for apparently controlling his brain's electrical output: with the EEG's electrodes affixed to his skull, he attempts to produce electrical signals between 10 and 12 Hz by following a protocol that Dewan had been developing to aid pilots in the control of their brain waves, involving a precise rhythmic and sequential opening and closing of the eyes.[7] Despite this protocol having nothing to do with sound, music, or electronic audio signal, the author of this article casually associates each of Lucier's eyes to the left and right channels of a stereo audio signal, implying that such a signal emanated directly from Lucier's brain. These "brain wave" signals, as this author characterizes them, "epitomize the natural or 'live' impulses which excite the frontiersman of the art."[8] Unlike the communist mind, Lucier's mind apparently had more options for input and output. But like the communist mind, Lucier's mind was still a black box, a complex, opaque system whose behavior could only be observed at its output. And not only was Lucier's mind a black box, it was also a cybernetic black box; its output was fed back into its input, effecting self-control.

The technologies that promised to allow Lucier to explore such apparent "natural or 'live impulses'" in the 1960s were numerous, each emerging from the military-industrial-academic complex in which Lucier was situated in the greater Boston area. Beginning in 1963, Lucier was employed by the recently founded Brandeis University as the director of its choir and its electronic music studio; Brandeis's geographical location gave Lucier proximity to military research undertaken at the Air Force Cambridge Research Center (AFCRC), Sylvania, and other nearby contractors. Throughout the 1960s, in addition to Dewan's EEG, Lucier also utilized very low frequency (VLF) radio antennas, Sylvania's vocoder, and a unique device called a Sondol (a portmanteau of "sonic dolphin"), developed by a nearby military contractor for the Office of Naval Research, as discussed later in this chapter.

Two major interpretations have dominated scholarly and artistic discussions of Lucier's early work with these devices. The first is that Lucier utilized these devices to tap into the natural, raw, or otherwise objective domain of "sound" itself, much like how Oliveros understood her work with oscillators and tape delay. This interpretation has been championed by many of Lucier's students; for example, in the words of the composer Daniel L. Wolf, "the music of Alvin Lucier doesn't begin with a musical convention or an abstract compositional notion—*an idea*—but begins instead with the nature of sound itself."[9] Articulated more robustly by the philosopher Christoph Cox, this interpretation holds that Lucier's music "is never about the signifier but always about the sonic real, sonic materiality itself."[10] Utilizing an idiosyncratic, heterodox interpretation of the psychoanalytic theory of Jacques Lacan, Cox claims that "Lucier's project is aimed at uncovering what undergirds the symbolic order but is disavowed by it: what Jacques Lacan called 'the real,' the perceptible plenitude of matter and nature."[11] For Cox, Lucier's early work with technologies such as the Sylvania vocoder "announces his commitment to a thoroughgoing naturalism or materialism,"[12] revealing what Cox elsewhere calls "sonic flux," in which "sound is immanent, differential, and ever in flux."[13] In this reading, "there is no strict division between subject and object, self and world, perception and substance—and this is not because [...] everything is mind or discourse, but because everything is nature or matter"; for Cox, "Lucier [...] draws no rigorous boundaries between the human, the animal, and the machine."[14]

The other common interpretation of Lucier's early electronic works, however, calls attention precisely to its reliance on historically and socially situated technologies, many of military origin, that reveal Lucier's supposed "naturalism" to be an artistically theatrical illusion. This approach has been undertaken by Volker Sträbel and and Wilm Thoben, for example, in their analysis of Lucier's *Music for Solo Performer*, the work that emerged from Lucier's encounter with Dewan's EEG apparatus. Sträbel and Thoben show how performances of this work are not revelations of natural sonic flux, but rather rely on "a subtle play of hidden communication between soloist and assistant" that ensures the piece achieves its desired theatrical effect.[15] Ultimately, for Sträbel and Thoben, Lucier's work "can be reduced to neither its method of sound production nor its sonic reality," since it is an "artistic production" rather than a scientific experiment.[16] This approach is echoed by the art historian Douglas Kahn, whose extensive research on Lucier's early work balances historical inquiry into the social and material bases for his work—that is, his entanglement with military technologists such as Dewan—and also assesses the aesthetic qualities of Lucier's work, locating it in histories of American experimental music that thematize the dynamics of energy and control.[17]

Both of these interpretations cite Lucier's early works from the 1960s that continued to be performed into subsequent decades—and for which

Lucier eventually created scores, many of which were published years after their respective works' premieres. Indeed, beginning in the mid-1970s, Lucier began to pare down his extensive work list from the 1960s and create scores for the works that remained, many of which were eventually published in the 1980 volume *Chambers*; these scores are frequently cited by scholars and also used by performers.[18] But in the crucial early years of Lucier's work in the Boston-area military-industrial-academic complex between 1963 and 1967, many other works were imagined, sketched, and even performed without surviving into the next decade. These works inhabited much more ambiguous spaces between scientific knowledge, technological exploration, and artistic production than the spaces in which Lucier's career would thrive from the 1970s until his passing in 2021. These early ambiguous spaces, which emerged from Lucier's specific encounters with and readings of American cybernetic and information-theoretical technoscience in the 1960s, played differentially with notions of human and nonhuman subjectivity, agency, embodiment, and performativity; they did not uniformly explore "sonic materiality," nor were they monolithically theatrical. Instead, they bear the traces of Lucier's encounters with and operationalizations of information theory, cybernetics, and computational metaphors that characterized American technoscience in the 1960s, producing ambiguous, tenuous, experimental, and fleeting glimpses as the many different relationships and boundaries that could be drawn between human, animal, and machine.

This chapter focuses on two major works that survived from this early period—1965's *Music for Solo Performer* and 1967's *Vespers*—and several related works that did not, but which Lucier saved in his archive. By closely following Lucier's negotiations between technoscience and music in these works, this chapter characterizes Lucier's early work neither as a revelation of natural sonic material nor as a theatrical reflection of cybernetic or information-theoretical principles. Rather, recuperating the ambiguity of these works, it argues—using a conceptual metaphor developed by feminist scholars of science and technology Trinh T. Min-Ha, Donna Haraway, and Karen Barad—that Lucier's work does not "reflect" extant ideas from cybernetics and information theory, but rather "diffracts" them, producing a relation of difference rather than similarity.[19] That is to say, if Lucier's early works suggest that the human being is in some way a cybernetic black box, an opaque processor or generator of informational signal, then that black box, unlike the speculative black boxes developed by Subotnick, Buchla, and Oliveros, was radically underdetermined. In this way, Lucier's black-boxed musicality comes closer to Sun Ra's profound opacity in its refusal to be wholly legible within any earthly scientific (or musical) paradigm. In Lucier's early works, the human performer does not retain complete authorial control, as it does in Subotnick's work; it is not an equal partner with an externalized computational model of the mind, as it was in Buchla's work; and it is not a wellspring for the refashioning of the self, as it is in Oliveros's work. Its subjectivity is not entirely abnegated, however, as discussed in the next chapter about Sun Ra's performative relationship with the Minimoog. In Lucier's early work, the black-boxed human performer produces an ambiguous, shimmering opacity, making differing possibilities emerge as Lucier explored the aesthetic, social, and poetic consequences of experimenting with the instruments of 1960s technoscience.

MUSIC FOR SOLO PERFORMER

Lucier's entanglement in the world of postwar technoscience came at a moment of personal creative crisis. In the early 1960s, after nearly fifteen years of traditional musical training, Lucier reflected that he was "at a point in [his] life when [he] didn't have any good ideas," and that he "certainly didn't want to compose instrumental music."[20] The rapidly developing field of what came to be called experimental music offered a way out. While in Europe on a Fulbright scholarship to nominally study at the Conservatorio Santa Cecilia in Rome, Lucier collaborated with fellow American ex-pat composer and pianist Frederic Rzewski on a concert series, worked at the Studio di Fonologica in Milan, and traveled to Darmstadt, West Germany, to attend a seminar with the pianist and composer David Tudor.[21] Upon his return to America in 1963, Lucier was hired by Brandeis University as the director of its choir and its newly founded electronic music studio.[22] Throughout the mid-1960s, Lucier audited classes at Brandeis on acoustics, physics, and computer programming and read widely in contemporary scientific fields. He took voracious notes that often contained the seeds for compositional ideas alongside detailed descriptions of phenomena in acoustics, communications, and ethology.[23]

Soon after Lucier's arrival at Brandeis, his chance encounter with Dewan, an old Yale classmate, led him to develop what would eventually be called *Music for Solo Performer*, perhaps the best known of his works from this early period. As Douglas Kahn has thoroughly documented, at the time of this meeting, Dewan had already been conducting research for several years on the electrical activity of the brain, commonly known as *brain waves*, at the AFCRC.[24] Dewan's research had been encouraged by none other than Norbert Wiener, one of the founding theorists of cybernetics, whom Dewan had met in 1960; under Wiener's influence, Dewan thought that using an EEG to measure the power spectra of the human brain—how its overall electrical activity was distributed among different frequency components—might create a set of data that could contribute to an informational model of human consciousness.[25]

The AFCRC, however, was interested in a much more acute problem. Air Force helicopter pilots had been experiencing episodes of disorientation and even unconsciousness while flying, which was thought to somehow relate to the frequency of the rotation of their helicopters' rotors; Dewan was assigned to solve this problem by researching the relationship between visual stimuli and consciousness. In the course of this research, Dewan developed a protocol to apparently control the electrical output of his brain (via scalp-mounted electrodes) by entering into what he called a "meditative state" and controlling the aim and focus of his eyes, as well as the opening and closing of his eyelids; this electrical output, when it fell within a certain frequency range, could be used as an informational electronic signal. A 1967 article published in *Nature*, still widely cited, describes a protocol for using the EEG apparatus to "train" a human to control their "brain waves" through audible and visual feedback, with enough precision that it could be possible to use the brain to send encoded messages. (Not surprisingly, the word used as an example of such a message in Dewan's experiment was *cybernetics*.)[26] A 1964 news segment aired on Dewan's local CBS affiliate in Boston showed both Dewan and an unnamed assistant demonstrating his apparatus, switching a lamp and a television set on and off, and "sending morse code messages" via chart recorder.[27]

Throughout 1965, Lucier experimented with Dewan's equipment at Brandeis, attempting to put himself in a meditative state by closing his eyes and executing Dewan's oculomotor protocol, as demonstrated two years later for the *Sylvania Scanner*. During a visit and concert at Brandeis by John Cage in 1965, Cage insisted that Lucier put a work on the program; Lucier demurred, saying he didn't have anything except for his experimental brain wave apparatus, which wasn't working. As Lucier recalls, Cage's response was that it didn't matter if it didn't work.[28] Lucier premiered his "brain wave piece" on May 5, 1965, at the Rose Art Museum at Brandeis, assisted by Cage and Anthony Gnazzo (who would soon thereafter be tapped to briefly direct the SFTMC at Mills College). For this piece, which was not yet entitled *Music for Solo Performer*, Lucier sat on stage with the electrodes of Dewan's EEG affixed to his scalp; the electrical potentials detected by the electrodes were transduced, filtered, gated, and impedance-matched before being sent to amplifiers and then directly to loudspeakers placed on percussion instruments. The subsonic signals produced by Lucier's brain were not directly "sonified," but rather used to physically vibrate the loudspeaker cones, which in turn activated the acoustic resonances of the percussion instruments on which they were placed. As Lucier remembers, this performance was not well received: "Two of my Brandeis faculty colleagues were there, too. During the performance one of them pretended to be asleep while the other gave him a hotfoot."[29]

Despite its harsh reception from Lucier's colleagues, *Music for Solo Performer* has long been cited as an early demonstration of cybernetic *biofeedback*, of the potential for a human being or biological process, understood as a self-regulating system, to control itself through its own informational activity—in this case, its brain waves. Its performance presents an apparent transduction of the solo performer's brain waves to sound, which are then "fed back" to the performer's auditory sensory apparatus, in turn controlling the mental state that determines the frequencies of the brain waves themselves (even though Dewan's original experimental control protocol was strictly oculomotor and did not involve auditory sensing at all). This mirrors Dewan's explicit involvement with, and invocation of, cybernetics in his own scientific research. Because of Dewan's proximity to Wiener and Lucier's proximity to Cage, Kahn argues that "Music for Solo Performer can thus be understood as a manifestation of cybernetics within music, a meeting of Wiener and Cage, one step removed."[30]

To argue for a reading of *Music for Solo Performer* as a reflection of cybernetic principles, Kahn cites performance materials for *Music for Solo Performer* created in 1972, which place the human being in the center of a block diagram that specifies the piece. Thus, Kahn argues, "*Music for Solo Performer*, in our nineteenth-century jargon, places things that are usually not regarded as technological in-series or in-circuit."[31] The historian of science Andrew Pickering similarly cites the work as an example of what he calls "ontological theater," a performance that stages a nonmodern ontology by presenting "a vision of the world as a place of continuing interlinked performances" while also serving as an example of how such performances might be manifested through man-made technologies.[32] For Pickering, Lucier's piece is "a reciprocal and open-ended interplay between the performer and the performance, with each both stimulating and interfering with the other—a kind of reciprocal steersmanship," a rough English translation of Wiener's neologism "cybernetics."[33]

Both Kahn and Pickering understand *Music for Solo Performer* as a system of flowing signal, which Kahn points out is a kind of energy: "[E]nergy becomes the great equalizer, circulating through a transductive flow in and out of electrical/ electromagnetic, acoustical and electrochemical energies, in and out of signals and sounds, perceptible and imperceptible states, an arrhythmia of speeds, traversing the distances of electronic circuits, performance space, physiological and neurological pathways."[34]

It is this seamless flow of energy and information between matter that is valorized not only in Lucier's own descriptions of the piece from 1972 onward, but also in citations of *Music for Solo Performer* as a foundational work in the field of scientific data sonification. Writing in the *Oxford Handbook of Computer Music*, for example, Atau Tanaka locates *Music for Solo Performer* as "the seminal work in [analog biofeedback systems]," using the brain itself as a "biosignal instrument."[35] Similarly, the composer Andrew Dewar writes that the work "place[s] the EEG into a musical context *without modification*," reframing a medical practice as a musical one.[36] In these readings of *Music for Solo Performer*—and indeed in Lucier's score and performance materials from 1972—the work is characterized as reframing the apparently natural phenomenon of human brain waves and acting as a reflection of the scientific discourse of cybernetics. This allows, and indeed invites, a strong sonic materialist reading of the work, positing the brain as the generator of transducable electronic signals, a potentially "natural" electronic musical instrument. This is the position articulated not only by the *Sylvania Scanner* in 1967 but later by Lucier himself. In 1980, he recalled a revelation he had while working with Dewan's equipment that "electronics comes from your brain, from inside every person, that every person has a little electronic studio inside his or her brain."[37]

Lucier's later position on the relationship between human musicality, the human brain, and electricity relies on the idea that electronic brain waves are natural phenomena that have always existed and were simply waiting to be discovered. But as the historian of science Melissa Littlefield has shown, conceptualizing the brain as an electrical system that outputs brain waves only emerged in the 1920s with the development of instruments of early electroencephalography. Such "visualizations of the brain's electrical activity," like the oscilloscopes and paper chart recorders of Dewan's apparatus, "are [both] products and producers of brain wave ideologies"—ideologies that valorize the "(electrical) *potential* of our brains to produce measurable states of mind and be open to intervention."[38] As Littlefield argues, taking brain waves for granted produces an understanding of the human subject that imagines "psychological or behavioral ontologies as physiological substrates."[39]

Brain wave ideologies enable a cybernetic reading of *Music for Solo Performer* by positing Lucier's brain as a transceiver of electronic signals that could cybernetically control itself through sensing those same signals as sound; they also enable a materialist reading of the work by imagining that "the brain itself is an electrical generator and a musical instrument in its own right," as Christoph Cox writes of the work.[40] For interpreters such as Cox, understanding brain waves as scientific, objective, or otherwise natural phenomena produces a conceptual equivalence between the fluctuating electrical potentials of the brain and the fluctuation of pressure waves in air, of sound as a fungible signal that seamlessly travels between acoustic and psychoacoustic substrates. But attending to Lucier's actual practices with the technologies that could produce such fluctuations of electrical potential in the 1960s complicates this reading. Rather than his work being "never about the signifier but always about the sonic real, sonic materiality itself," as Cox claims, Lucier's work from this period—especially in sketches and other works produced contemporaneously with *Music for Solo Performer*—was explicitly interested in signifiers and the poetic ambiguities that could be explored between such signifiers and a concept of nature that was constantly being renegotiated.[41]

SKETCHES, 1965–1966: SIGNAL OR SIGNATURE?

Virtually all extant scholarship on *Music for Solo Performer* references materials that Lucier began to publish after its 1965 premiere. These include a text published in the *Fylkingen Bulletin International* in 1967 that describes the work; performance materials prepared in 1972 to allow the work to be performed by people other than Lucier, which included the block diagram already discussed; and a text score authored by Lucier and published in *Chambers* in 1980.[42] But in the year of Lucier's initial encounter with Dewan's EEG equipment in 1965, and indeed continuing into 1966, his sketches and notebooks suggest that he was not particularly interested in the idea that the brain could be a generator and processor of electronic signal. Indeed, the fact that Lucier was reluctant to perform with the EEG apparatus, and that it was only at Cage's insistence that the unnamed, unfinished piece was premiered, suggests that in 1965 this work was still speculative—like many of the works that Lucier saved in his archive, but which were never completed, or in some cases were intentionally dropped from the composer's work list in the 1970s. These unfinished, speculative, and discarded works tell a far different story about Lucier's relationship with contemporary technoscience. Rather than his musical work revealing a Coxean "sonic materiality," with the performing human acting as a black box that processes a naturalized concept of sound, these works poetically speculate and generate sound that is both natural and cultural, historically situated, and socially grounded.

One sketch, undated and untitled but collected with several others from 1965 and 1966, begins with the instructions: "Find, collect and use any man-made or natural physical objects or visible signs as scores for music."[43] While using an ostensibly natural object as an abstract graphic score for musical performance might suggest a Cagean interest in aleatoric indeterminacy, such as Cage's use of a star chart to compose *Atlas Ecliticalis* (1961–1962), Lucier's sketch lists nearly fifty "man-made or natural physical objects or visible signs," many of which rhyme and share similar poetic meter:

> Eagle feathers / door jams / razor clams / radiator pipes / flying kites / pressed shale / foreign mail / canvas sails / wet leaves / dry gulches / strawberry jam / cinder blocks / grandfather clocks / pekoe tea / rough seas / trilobites / fossils / footprints / spilled ink / bee's nests / squashes / galoshes / weather maps / jockey's caps / wooden slats / aerobats / puddings / dumplings / lettuce stalks / Kleenex boxes / stains / scratches / glacial state / shopping lists / fingerprints / palms / reiner heads [?] / meandering streams / tire marks / wind on water / sand dunes / waves on beaches / canoes on lakes / falling stars / crashed cars.[44]

Although some of the objects in this list are traces or artifacts of weather patterns or other ostensibly natural phenomena, just as many are firmly culturally and historically situated in the mid-twentieth century, creating a free-associative poetic sketch of Lucier's life in New England. This work remains firmly in the realm of the symbolic, rather than employing the notion of sound as a phenomenon separated from the human world.

Collected along with this sketch in Lucier's archive is another that bears the title *Signatures*; a transcription of the score appears in figure 16. *Signatures* instructs performers to understand the "principles" of natural processes and to make sounds according to those principles without specifying how. By drawing a clear distinction between "signal" and "signature," and by editing this sketch to change the verbs "make and leave" to "write," Lucier valorizes the symbolic inscription of such signatures over any potential signal that they produce. This sketch implies that a performer of the work might not even produce sound; they are instructed to "write new signatures (for the next player)" before they are instructed to "read the signatures," although such readings could easily be silent. Lucier's work plays with the illegibility of these signatures; iteratively written and read, they behave as opaque generators of an ambiguously poetic musical performance.

While *Signatures* begins from inscriptions "written" by natural processes, another work from 1965 seems, at first, to investigate *signal* itself. This work, variously titled *Music for Amplified Lip* or *Composition for*

Signatures

Look up word in dictionary -
Everything / everyone has and leaves its / their signature: fossils, footprints, shells, hands, etc.

<u>Sig</u> — <u>Nature</u>

        write
all visible things ~~are in the process of producing and leaving~~ their signatures.

Difference between SIGNAL + SIGNATURE.

Search and collect signatures ~~from~~ from the natural world.
Discover how these signatures were made.
Use these principles to make sounds isolated or together.
get groups of people together according to these principles.
    write
Make and leave ~~onto~~ new signatures. (for the next players.)
Read the signatures.

Look for and collect signatures

Utilize principles of processes -
 stratification of rock
 markings on shells
 cirques

Figure 16. Transcription of Signatures score. From ALP, box 41, folder 4.

Amplified Lip, is listed in a 1966 curriculum vitae as having been composed before *Music for Solo Performer* in 1965.[45] The score for *Lip* calls for two contact microphones to be placed on a performer's lower lip; the signals from those microphones are preamplified and sent to one to ten assistants who are positioned in a semicircle around the performer, "equipped with modifiers, tape recorders, amplifiers and speakers in any combination."[46] Although similar to contemporaneous performances of John Cage's 1962 work *0'00"*, which the composer had performed by amplifying the sound of himself drinking through a contact microphone affixed to his throat (and which had been performed on the same concert program as Lucier's brain wave piece in 1965), *Lip* was much more technologically spectacular. In a typewritten draft of the score, Lucier wrote that "technical assistants are free to use as much of the solo performer's sounds as they wish" and underlined a handwritten note that "assistants are free to exchange material."[47] In this draft, the word *sound* is crossed out and replaced with *input*—suggesting that what was circulating in this piece was not sound itself, but rather a signal carrying sonic information.

Unlike *Music for Solo Performer*, which emerged after *Music for Amplified Lip* and was not initially scored, *Lip* was composed with a graphic score of gestural material for the performer to execute, with an elaborate notational key to Lucier's "general vocabulary" that included detailed instructions for lip smacks, tongue clicks, snapping the tongue against the palette, a Bronx cheer, sucking, blowing, and thirteen other specific actions.[48] The signal generated from these actions would be fed into an elaborate array of communications equipment that Lucier intended to spectacularize during the work's performance. A few weeks before its 1966 premiere, Lucier wrote to Gordon Mumma, who was helping to organize the concert: "I guess I'm really thinking of *Lip* which to be good, has to be very complex. The technicians in the piece must be responsible for at least two modifications of the central sound source. I want the loop machine (Schaeffer/Close Study Unit), a reverb unit, a filter and any other gadget you could bring. David [Behrman] is making a vibrato simulator which should work just fine. How about your Dallas box? I would also want to use pre-recorded tapes and loops as a safety factor."[49]

Although *Lip* involved the production of "live" signal from the human body, Lucier's approach to the work's performance, which included elaborate preparatory material, a graphic score, and the suggestion that prerecorded signal could be used, complicates a valorization of signal over signature, to use Lucier's terms. Even if the signal were generated live, the technicity of its mediation through communications technology was spectacularized, and the signal would certainly not be heard without some sort of modification that would have been clearly apparent to the audience; the score calls for the "technical assistants" in this work to sit in a semicircle facing the "solo performer" on stage so that their devices could be seen. Lucier's sketch also underlines the importance of the visual spectacle of the assistants "exchanging material," which would have meant the laborious processes of winding and rewinding 1/4-inch audiotape, threading and unthreading it through different machines, and possibly even cutting and splicing in real time. The staging of *Lip* starkly contrasts performances of *Solo Performer*, in which the solo performer is alone on stage, and the technical assistants—who, as Sträbel and Thonen have shown, are essential for the work's functioning—are hidden. Likewise, the sound of its performance as described in Lucier's letter would be strongly colored by the chaotic mixture of the various modifications effected by tape looping, filtering, and reverberation—a far cry from the soberly intermittent and ostensibly indexical thuds of loudspeaker cones on percussion instruments.[50]

Because it was composed immediately before *Music for Solo Performer* and eventually dropped from Lucier's work lists in the 1970s, it is tempting to read *Lip* as a failed preliminary sketch for *Solo Performer*—which Lucier came to describe as early as November 1966 as "my first work [. . .] exploring with sophisticated sensing devices such as vibration pickups and high gain amplifiers."[51] But the fact that *Lip* was premiered a year *after Solo Performer*, and that Lucier evidently planned its performance to emphasize the materiality, technicity, and sociality of the "signal" coming from the performer's lip, complicates this reading. *Lip* suggests that Lucier was not wedded to one specific understanding of the relationships between human bodies, human consciousness, technical objects, sound, and informational signal in the 1960s. And indeed, another work from this same period, 1967's *Vespers*, evidences a similar ambiguity, showing the diffractive and ambiguous approach that Lucier took toward incorporating the concepts and technologies of contemporary technoscience into musical practice.

LUCIER AND THE HUMAN BIOCOMPUTER

Lucier has often framed his early electronic works as emerging from encounters with specific "sophisticated sensing devices," as he put it in 1966, which promised to transduce signal from various sources—human brains, the ionosphere, or even buildings—into sound. In a 1980 interview with Douglas Simon, Lucier traced his 1967 work *Vespers* back to an encounter with a device called a Sondol, a handheld instrument that emitted a narrow directional beam of clicking sounds that would reflect off various surfaces, ostensibly allowing its human operator to learn how to locate themselves in space.[52] But in that same interview, Lucier also stated that "like most of my pieces I thought about it for a long time before I actually made the final realization."[53] Although Lucier never publicly detailed the thought that went into the development of *Vespers* before it was realized with the Sondol, his archive suggests that much of that thought emerged from his encounter with the ethological and philosophical thought of the neurophysiologist John C. Lilly, whose 1967 book *The Mind of the Dolphin: A Nonhuman Intelligence* likely instigated Lucier's initial kernel of an idea for a work relating to human-animal communication.

As discussed earlier in this chapter, Lucier was a frequent auditor of and voracious notetaker in science courses at Brandeis; his mid-1960s notebooks contain notes on topics ranging from surface wave behavior to electrical engineering to computer programming in FORTRAN. A series of four pages ripped out of one such notebook, undated but likely from 1967, contains detailed notes on Lilly's 1967 book *The Mind of the Dolphin*.[54] By that year, Lilly had gained a certain amount of fame and notoriety for his eccentric experiments, theories, and writings about interspecies communication—specifically between humans and *Tursiops truncatus*, the common bottlenose dolphin. Written throughout a long period of military sponsored cetacean research that had been nationally popularized by media such as the *Flipper* film and television franchise, Lilly's books regaled readers with tales of fantastical experiments, conducted since the 1950s, which suggested that the auditory and vocal behaviors of bottlenose dolphins offered evidence of a nonhuman intelligent mind.

Lucier's interest in Lilly may have emerged from the fact that Lilly's experimental practices relied extensively on the same cutting-edge audio technologies that Lucier also used in his own works. Lilly used novel hydrophones, specialized loudspeakers, and magnetic tape to record and play back the "signals" generated by *Tursiops truncatus* in various experimental protocols. Indeed, as John Durham Peters has shown, after initially perceiving potential interspecies communication using only his own ears, magnetic audiotape became the *"condition sine qua non* of Lilly's cetacean research."[55] Taped sounds of experimental interactions with dolphins, transformed through time-axis manipulation and visualized through spectrographic analysis, became the substrate for a basic assumption that undergirded much of Lilly's research: that sound was a medium of communication, and that "communication is the exchange of information between two or more minds."[56]

While Lilly's theorizations about communication were shaped by the material and technological affordances of audio technologies, his understanding of "mind" was shaped through the basic principles, materials, and techniques of computing. Lilly considered *mind* to be an emergent property of a brain "residing in a body detectable by the senses or by the senses aided by special apparatus."[57] As he hypothesized in *The Mind of the Dolphin*, and as Lucier summarized in his own notes, "the absolute size of a mammalian brain determines its computing capability and the size of its storage (memory)," and that "the theories (programs and metaprograms) stored in one's self operate the way a stored program in a modern computer operates."[58] Based on these assumptions, Lilly developed an idiosyncratic ethics for experimental praxis: "Thus, to test a given theory, one 'programs' himself with the as-complete-as-possible theory and joins the system under investigation as a participant computer operating 'on-line'. This operation of a computer 'on-line' is a new concept in computer technology. It means that instead of using punch cards or magnetic tape for feeding the information into the computer, the computer operates from data being generated in real time, right now."[59]

Not surprisingly, Lucier's notes on *The Mind of the Dolphin* focus on descriptions of the various physical properties of cetacean and human acoustic communication in air and water and on the physical differences between dolphin and human anatomy that would have consequences for the ways they produced and sensed sound. But these notes also suggest that he was just as interested in Lilly's computational theory of mind. In these notes, Lucier describes both humans and cetaceans as information-processing "computers" and focuses on what he summarized as "physical information theory": "higher signal, more physical info in, greater number of bits/sec."[60] Indeed, Lucier's notes on what he termed "on-line computational thinking" summarize a set of axiomatic imperatives that Lilly believed were necessary for experimental practice:[61] that "truth is sought"; "one streamlines himself for the program"; and that in the moment of experimentation, "one starts the almost endless process of creation of the pertinent 'software.'"[62] In order to seek truth, Lilly wrote, one must try to be like a computer, which would involve not only "get[ting] rid of excess emotional baggage," but also "hav[ing] a metaprogram which says 'create necessary subroutines, routines, programs, metaprograms, and models to find the truth as it exists in real systems in which one is interested.'"[63] Engaging in this kind of "on-line computational thinking" would lead to the discovery of truth via the "meta-programming" of what Lilly called the "human biocomputer." In more general terms, Lilly's experimental ethics effected an investigation into the human biocomputer as if it were a black box in Wiener's original 1948 definition of the term, an "as-yet-unanalyzed nonlinear system."[64]

In 1967, Lucier began to draft a research project proposal that contained specific references and similarities to Lilly's research program described in *The Mind of the Dolphin*. The title of Lucier's proposal, which went through many edits, was initially drafted as "The Communication with Non-human Intelligences by Means of a Computer-Generated Musical Composition," which echoed Lilly's exploration of "non-human intelligences."[65] In addition, Lucier's proposal sought "to learn to share in the extraordinary sound-sending and -receiving of the dolphin, *Tursiops truncatus*," the same species studied by Lilly.[66] Beyond the subject similarities to Lilly's research, Lucier's proposal also shared its fundamental premise that a human being was a "participant computer" whose mind could be "meta-programmed" through the use of novel sensory apparatuses in real time. Indeed, Lucier's proposal described a "computer-generated musical composition" without any mention of an electronic digital computer, implying that the human performers of the work would act as biological computers engaging in what Lucier, after Lilly, called "on-line computational thinking."

Lucier never publicly discussed this proposal, and indeed, if it was a proposal prepared in anticipation of seeking funding, it appears never to have been funded.[67] On one page of the proposal—which was evidently written over the course of weeks or months, in different inks, on different torn-out pages, and with multiple revisions of various sections—Lucier calls for the purchase of a device called an "Environ-Ears Binaural recording system, available on order from Listening, Inc." Although Lucier eventually did use a similar binaural microphone system in 1972 to record the work that emerged from this proposal, it was a different piece of equipment from the same manufacturer that more acutely sparked his interest: a device called a Sondol, a portmanteau of "sonic dolphin," which promised to give humans the ability to learn the skill of echolocation. While Lucier later characterized the Sondol as a scientific instrument—and as such, "free of content" (i.e., a "social or political message," as he wrote in 1984)—mapping the Sondol's origins in the Boston area military-industrial-academic complex reveals a layer of social and political meaning that would have been difficult to ignore, and as I will show, was emphatically not ignored in early performances of the work.[68]

TOWARD *VESPERS*

Listening, Inc., based in Arlington, Massachusetts, was an enterprise situated between the worlds of military cetacean research, popular parascience, and overt pseudoscience. Its CEO, Dr. Dwight Wayne Batteau, was a US Army Signal Corps veteran and a mechanical engineering professor at Tufts who was also an early adopter of Scientology and a regular contributor to the science fiction periodical *Analog Science Fiction and Fact*.[69] In addition to employing several of Batteau's own Scientological initiates, Listening, Inc. also employed a local autodidact named G. Patrick Flanagan, who as a teenager developed a device he called the Neurophone that he claimed could transmit audio signals directly into a human brain.[70] In 1966 Batteau secured a contract from the US Naval Ordinance Test Station (USNOTS) to study "Man/Dolphin Communication" and employed Flanagan, among others, to design instruments for his research.[71] According to its final report to the USNOTS prepared in December 1967, Listening, Inc.

spent its funding on the development of two computing devices: a "Man to Dolphin Translator," which would encode human speech as frequency-modulated "whistles" transmitted to dolphins via underwater speakers, and a "Multi-Spectrum Analyzer," which would record dolphin utterances via hydrophones and provide spectrographic analysis on paper.[72]

But in an earlier report—this one published in the July 1967 issue of *Analog Science Fiction and Fact*—Listening, Inc. advertised a very different set of instruments developed with USNOTS funding.[73] Among these were Flanagan's Neurophone (in an updated design), a series of large metal "ears" designed for listening in air and water, and the Sondol, which would apparently bestow upon humankind the cetacean ability to echolocate. These instruments were designed using a computational model of human sensation, cognition, and consciousness similar to Lilly's; a photograph of Flanagan speaking into a microphone next to a steel "ear" roughly the size of a human head bears the caption: "Ear is an array designed to be used in conjunction with the human computer."[74] Like the "Ear," the Sondol was an instrument designed to be used by the "human computer," providing it with a stream of real-time information that would theoretically allow it to locate itself in space (see figure 17).

Lucier remembers meeting Flanagan by chance when he went with the visual artist Mary Lucier, then his spouse, to find an empty garage for Mary to use for her sculpture practice. Flanagan told Lucier about Listening, Inc.'s research and loaned him a prototype of the Sondol, along with his older instrument, the Neurophone.[75] Although Lucier never made work with the Neurophone, the Sondol seems to have offered an elegant solution for his earlier research proposal for "the communication with non-human intelligences by means of a computer-generated musical composition." Indeed, the Sondol became central to the development of a new work that emerged from this proposal, which Lucier eventually titled *Vespers*.[76]

The Sondol was a small, handheld device developed by Listening, Inc. employee Stephen Moshier, one of Batteau's initiates into the Church of Scientology.[77] Its patent describes it as "projecting a beam of high frequency, staccato like, acoustic pulses which, when directed toward a distant object capable of reflecting an echo, will enable the user to determine the distance and character of the object by interpreting the quality of the echo."[78] The Sondol's human user could theoretically determine the distance of an object by aiming it at that object and varying the rate of its

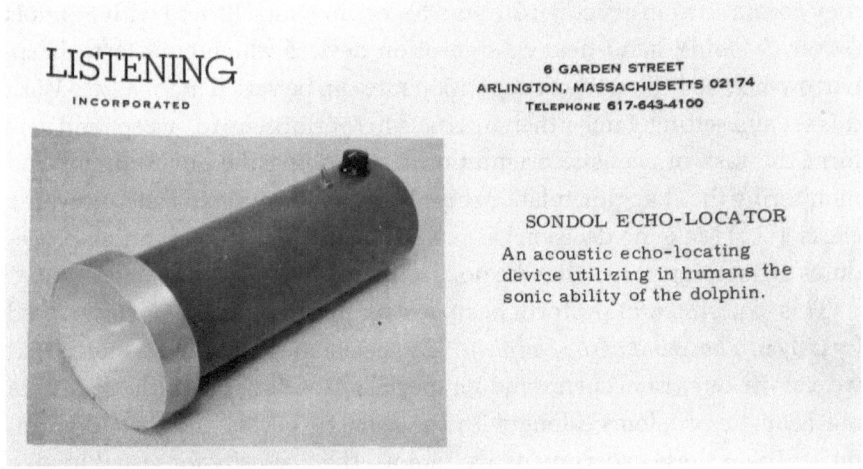

Figure 17. Brochure for the Sondol produced by Listening, Inc., 1967.

outgoing pulses until a particular (but unspecified) rate of reflected pulses was perceived. This simple experimental protocol for human echolocation in air does not match the complexity of the process by which dolphins locate themselves and objects in water, as described in Lilly's work; Lucier's close reading of Lilly would suggest that he would have understood this discrepancy.[79] Yet the simplicity of the Sondol promised Lucier an easy way to achieve the objective he articulated in his research project proposal: to "learn to share in the extraordinary sound-sensing and -receiving [capacities] of the dolphin, *Tursiops truncatus*" by changing the sensory apparatus of the human performer through audio technology.

With the Sondol, Lucier began to develop an experimental protocol that not only echoed Lilly's empirical ethological research into cetaceans but also enacted what Bruce Clarke has called Lilly's "experimental performance science," "in which [...] the experimenter self-fashions himself not just as a participant observer but as a computational device coupled into the system to be observed, open to on-line learning and real-time reprogramming."[80] Although Lucier would not write a score for this work until after its first performance in 1968, multiple drafts of performance scores remain in his archive, variously dated 1967 and 1968.[81] Many of these drafts begin with the instruction to "play in any dark space indoors, outdoors, or underwater; in dimly-lit spaces wear dark glasses and in lighted space wear blindfolds";

they continue to instruct performers to "equip yourself[ves] with Sondols (*son*ar *dol*phin), hand-held echo-location devices which emit fast, sharp, narrow-beamed clicks whose repetition rate can be varied manually."[82] With this set and setting, Lucier then instructs his performers to "accept and perform the task of acoustic orientation by scanning the environment and monitoring the changing relationships between the outgoing and returning clicks. [. . .] Make no decisions as to speed, direction, and scanning procedures of outgoing pulses that are not useful in the process of echolocating."[83]

This experimental protocol bears a striking similarity to one described by Lilly in *The Mind of the Dolphin*: "By special means it can be shown that we can use our visual cortex and its special characteristics in the service of our acoustic problems (along with the acoustic cortex and its special inputs). To do these experiments we remove the necessity for visual inputs, i.e. by closing our eyes or darkening the room. With proper techniques (hypnosis, drugs, etc.) we can program or compute an 'acoustic space' into the 'visual space.' For example, the blind may use the general purpose nature of the cortex in extending the 'acoustic space' into the 'visual space' to circumvent complex objects."[84]

Whereas Lilly specifies the "special techniques" of hypnosis or drugs to change the conscious state of the human experimenter, Lucier proposes that such a change could be effected through the use of the Sondols and "on-line computational thinking." Once his human performers are in this state, Lucier explicitly instructs them to avoid aesthetic decisions and instead only make decisions that will be useful for the experimental goal of echolocation. In this way, like Lilly's experimental protocol, Lucier's score treats sound not as an aesthetic object but as a real-time data stream to be studied. Indeed, several drafts of the score describe "sounds used as messengers, which, when sent out into the environment, return as echoes carrying information as to the shape, size, and substance of that environment and the objects in it."[85]

VESPERS: SIGNAL OR SIGNATURE?

In all of the published material relating to *Vespers*, sound is posited as an informational signal that, through the "on-line computational thinking" specified in the experimental protocol/score, might become decipherable by the human biocomputer. And indeed, when the work is performed today, performers often undertake good faith attempts to achieve such a feat—even in cases where there are few objects from which sound can reflect, such as a 2016 outdoor performance in the Trans-Pecos desert near Marfa, Texas. Despite the near impossibility of such a task, interpreters such as Cox often still describe Lucier's works such as *Vespers* as articulating a materialist sonic "naturalism" that cuts between bodies, species, and technologies; in a 2018 book, Cox maintains that "*Vespers* employs mechanical devices to endow performers with the sensory capacities of bats and dolphins."[86] However, as in the case for *Music for Amplified Lip*, Lucier's notes and correspondence surrounding the work's premiere in 1968 and a separate performance of the work in 1971 suggest a very different understanding of the role of sound in this work. These early performances valorize the social and cultural meaning of the work's performance over its scientific, ethological sheen.

For the work's premiere at the February 1968 ONCE Festival in Ann Arbor, Michigan, Lucier speculated about ways to provide a "visual analog" to "acoustic orientation by means of echolocation," the subtitle given to the piece in its preparatory notes.[87] These notes speculate about the use of "ultrasonic Sondols" (which did not exist) and a series of wire obstacles much like the experimental apparatuses used to study echolocation in bats employed by the ethologist Donald R. Griffin, whose 1958 book *Listening in the Dark* is referenced in the post facto score for *Vespers*.[88] Since such ultrasonic devices would not be audible to audience members, Lucier wondered if they could be ring modulated with an oscillator set to a single frequency, akin to the process of heterodyning in radio wherein high-frequency radio signal is transposed down into the audible range for humans. But alongside speculations about such modulation in his notebook, Lucier also wondered if these ultrasonic signals could be intermodulated with a "night radio broadcast" of baseball or some other "suitable input," which would eliminate any illusion that the work presented audience members with an indexical sonic trace of a natural phenomenon.

Lucier's speculative "visual analogs" to echolocation also firmly characterize this performance not as a presentation of natural sound, but as a natural-cultural representation of American life in 1968. Proposed "analogs" included a blind person tapping with a cane and illuminated by a strobe light, a fog horn, and performers wearing tap dancing shoes. Most spectacularly, Lucier wondered if "a very large game of squash could go on and off during the performance. Doubles. Lights out, balls visible and whizzing around and over audience. Should be frightening like bats getting caught in your hair. Net over audience or glass-enclosed squash court. [. . .] Instead of squash which could be dangerous we could use ping pong balls and paddles. Could a game of ping pong/squash be played in the dark? That is, could the players locate the ball by sound?"[89]

Although Lucier listed an "infra-red lighting setup for dark squash game" in his notes as a project to be completed prior to the work's premiere, he eventually settled on a much simpler activity to illustrate the phenomenon he was interested in exploring by simply placing obstacles in a room and instructing performers to navigate around them using the Sondols. This would not only give the performance concrete spatial boundaries, but it would also prompt the performers to inhabit the subjectivity of scientific test animals: "Perhaps the performers should have to avoid obstacles—chairs, people, etc.—by means of echolocation. Scan and avoid. Performers should be in a similar position to bats and be forced to suffer their tragedies."[90] To suffer the tragedies of experimental test animals, or to be frightened by a game of profoundly dark ultrasonic squash in which the ring-modulated signal of whizzing squash balls would be mixed with night broadcasts of local baseball games, is a far cry from the scientistic, objective language that made it into the work's eventual score.

Another performance from before the concretization of the work's score also highlights the impossibility of the task at hand and the valorization of the opaque poiesis of the image of humans attempting to echolocate in front of an audience. In a 1971 performance of the work filmed by a West German television crew for inclusion in a documentary about "new music" in New York, a group of performers including Alvin Lucier, Mary Lucier, David Tudor, and M. C. Richards walk slowly in a wooded area in the Gate Hill Cooperative in Stony Point, New York, slowly panning Sondols from left to right and constantly varying the rate of their instruments' pulsations.[91] Although Tudor is shown aiming his Sondol down at a nearby rock, which might have generated a perceptible echo, Alvin Lucier, Mary

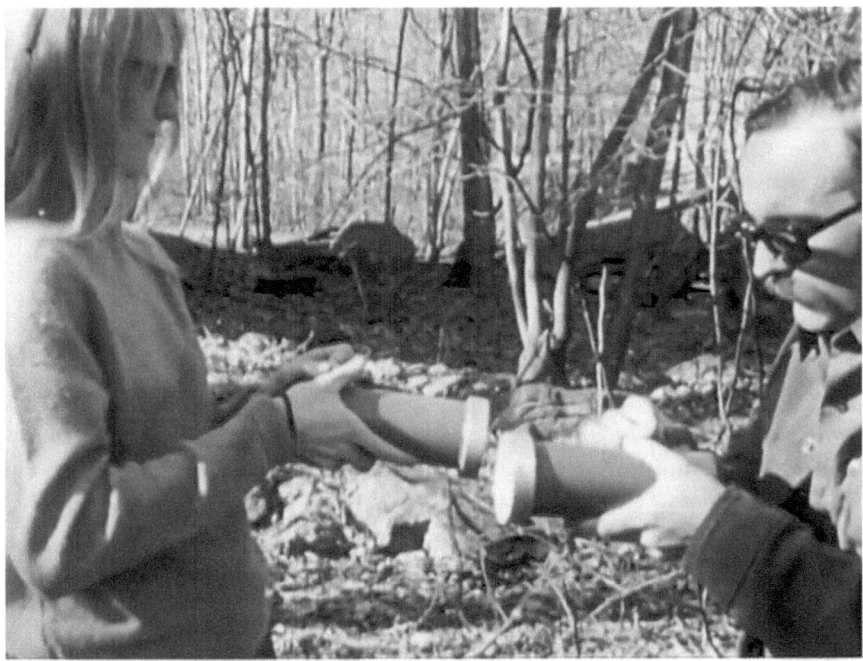

Figure 18. Mary and Alvin Lucier aim their Sondols at each other in a 1971 performance at the Gate Hill Cooperative in Stony Point, NY. From the film *New Music: Sounds and Voices from the Avant-Garde New York 1971* (2010), directed by Hans G. Helms, produced by Michael Blackwood Productions.

Lucier, and Richards all simply aim their Sondols straight ahead through the bare late winter trees (see figure 18). Far from echolocating themselves in space, the performers are seen moving toward each other in gendered pairs, with male and female performers slowly gravitating toward each other, manually increasing the rate of their Sondols' pulsations as they approach. This performance highlights the semiotic ambiguity of the supposed "on-line computational thinking" that Lucier imagined might take place while using the Sondol and also illustrates the near impossibility of its supposed ethological task.

An undated notebook page likely from the early 1970s, written in pencil, suggests that Lucier was still working through how to think about this work well into its performance life:

Vespers
Real echoes
Direct imitation, without metaphor,
    Without metaphor, but with ambiguity.
Direct imitation without metaphor but with enough ambiguity to be interesting or to provide information.
Geographical displacement made possible a poetic displacement.
Or a stylistic displacement in which two-dimensional notation written notation is eliminated in favor of [blank.][92]

The ambiguity and poetic displacement that Lucier located in *Vespers* is also illustrated by a sketch for a contemporaneous work from this period, *Computer Piece*, which operationalizes the technoscientific discourse of the Lillean "biocomputer" in a way that more explicitly valorizes the symbolic over the supposed sonic "real."[93] In this sketch, Lucier imagines a room of variable dimensions with variable objects—"pictures, hanging wires, molding on walls"— in which a person would navigate with "pulses or 10,000 cps sine tones" identical to the sounds produced by the Sondol. By slowly scanning the room with Sondol-like pulses, performers would not locate themselves in space, but rather would translate the spatial "information" of the room into language: "Rooms could become words. [. . .] If you walked a room sonically long enough, words or other things would appear depending on what spiritual situation or process was operating."[94] Lucier's sketch for *Computer Piece* speculates the production of a string of words by and for the "participant computer" through the performance of online computational thinking, an objective "output" of the newly reprogrammed human biocomputer able to translate spatial information to semiotic information. The poetics of this string of words would reflect not only the physical characteristics of the room, but also its "spiritual situation." Even if *Computer Piece* is read as a "direct imitation without metaphor," it is not a direct imitation of the natural world, but rather of a speculative, "spiritual" world, as Lucier put it.

Not only does *Computer Piece* attempt to imitate the supernatural; the output of such an imitation would be fundamentally ambiguous. How, exactly, would rooms become words (or "other things")? Would the words be in an extant language or a new language? Would the words exhibit a grammar or cohere in some form of syntax? Perhaps this is why Lucier abandoned the sketch, but such ambiguities remain at the heart of much of his work from this period. *Computer Piece* offers an explicit articulation of Lucier's technopoetics: the ways in which our world and our existence changes with, and through, technology. Well beyond a "direct imitation" of nature, both *Computer Piece* and *Vespers* produce a strong, and generative, poetic ambiguity.

NO IDEAS BUT IN THINGS?

The differing possible identities, behaviors, and actions of human subjects performing Lucier's works from this crucial early period exhibit Lucier's diffraction of contemporary technoscience into a spectrum of possibilities, rather than a simple reflection of scientific concepts in music. These works played with materiality and immateriality, signal and signature, immediacy and representation, in exploratory ways that did not produce a singular strong claim about the relationships between humans, electricity, information, and sound. Yet the question of materiality remained central to Lucier's work in this period, and indeed throughout the rest of his career. Musing on the materiality of a later work, 1977's "Music on a Long Thin Wire," Lucier cited a phrase from William Carlos Williams's 1924 epic poem "Patterson": "In a wonderful poem, William Carlos Williams has the line, 'No ideas but in things.' The older I get the more I realize that that's the way I work. My ideas come out of the actual doing of the work. I don't think beforehand about a system or need to control or figure out tensions and weights. I do it and see what happens, and I accept or not."[95]

"No ideas but in things" became a motto for Lucier in his later years when he was asked to explain his work; it was used as the title for a 2013 biographical documentary, in which his student Daniel Wolff claimed that Lucier's work "begins [...] with the nature of sound itself," as discussed earlier in this chapter.[96] The film includes video documentation of performances of *Music for Solo Performer* and *Vespers*, about which Wolff claims that "Lucier's task, as a composer, is that of finding circumstances for sounds to largely determine their own shape in time as a part of a unique musical form."[97] For Wolff, sounds, as natural and independent agents, determine the form of Lucier's compositions, which merely serve as protocols that allow such ostensibly natural sounds to be heard by and for themselves. Wolff's position sets the stage for stronger "sonic materialist" claims about this same period of work, such as Cox's argument that not only do such works reveal "sonic materiality," but such sonic materiality is *real*, beyond the symbolic order of language.[98]

For Wolff and Cox, imagining sound to be *real* and nonsymbolic necessitates thinking about the devices that produce and perceive such sounds—such as Sondols or even human brains—as cybernetic black boxes. As Andrew Pickering has written, such objects refuse what he calls the "detour through knowledge" and instead concern themselves with "performance as performance, not as a pale shadow of representation."[99] It is for this reason that Pickering has cited Lucier's work as an example par excellence of cybernetic "ontological theater," which simultaneously performs a "nonmodern" ontology in which there is no distinction between nature and culture, and also allows us to imagine "the sort of endeavors that might go with a nonmodern imagining of the world."[100] Wolff, Cox, and Pickering essentially take Lucier's early works at their word, evaluating them based on scores and performances from the decade after they were developed, once all of the technical kinks had been worked out and Lucier's worklist had been purged of works such as *Signatures, Music for Amplified Lip, Computer Piece*, and the many sketches that decidedly positioned sound as a modern, cultural, symbolic phenomenon.

Taking Lucier's early sketches, notebooks, correspondence, and other evidence from his archive into account, however, shows how Lucier's relationship to information theory, cybernetics, and American technoscience writ large held a very different valence in the 1960s. Where Wolff, Cox, or Pickering might hear works such as *Music for Solo Performer* and *Vespers* as nonrepresentational, offering immediate access to the *real* world of sound, I instead hear them as works that offer a shimmering opacity: a radically ambiguous, and often playful, sense of unknowable possibility. Lucier certainly grappled with the relations between human bodies, human sensation, audio technologies, information theory and cybernetics, sound, and "nature" in the 1960s. But Lucier's archive shows that in this crucial moment, his work did not solely privilege sound as a "real" or "nonmodern" phenomenon that could be revealed through experimental practice. In the sketch *Signatures*, Lucier's differentiation between "signal" and "signature" explicitly differentiates between the real and the symbolic; *Music for Amplified Lip* and *Music for Solo Performer* respectively spectacularize and veil the materiality of the creation of signal from human bodies; the premiere of *Vespers* created "visual analogs" for cognitive behaviors across species; and *Computer Piece* remained firmly in the realm of the symbolic.

This is perhaps why contemporary interpreters such as the musicologist Bernhard Rietbrock have tempered claims about the sonic real in Lucier's works, arguing instead that "Lucier experienced the real as an aesthetic event precisely not pre-symbolically, but always as an equally singular and universal effect of the cracks and gaps of the symbolic."[101] But while Rietbrock argues for what he calls a "reflective experimental aesthetics" of Lucier's work, which is certainly warranted, privileging *sound* over the haptic, visual, and other performative dimensions of Lucier's early works—not to mention their specific social, political, and historical situatedness in Cold War America—would still erase how Lucier himself conceptualized these early works in the decade of their creation.

To defend Lucier against his devotees, so to speak, it is worth returning to the poem by William Carlos Williams, *Paterson* (1926), from which Lucier sourced his later motto, "no ideas but in things":

> Before the grass is out the people are out
> and bare twigs still whip the wind—
> when there is nothing, in the pause between
> snow and grass in the parks and at the street ends
> —Say it, no ideas but in things—
> nothing but the blank faces of the houses
> and cylindrical trees
> bent, forked by preconception and accident
> split, furrowed, creased, mottled, stained
> secret—into the body of the light—
>
> These are the ideas, savage and tender
> somewhat of the music, et cetera
> of Paterson, that great philosopher[102]

The "things" in this poem—houses with blank faces, cylindrical trees, pauses between snow and grass and streets—are not purely "natural" and are likewise not easily placed as "nonmodern" things such as the cybernetic entities and systems present in ontological theater. They are characterized by overlaps of human endeavor, architecture, climate, plants, and weather; Williams calls these "things" the "ideas, savage and tender / somewhat of the music, et cetera / of Paterson." Williams moves from things to ideas to music—not only music itself, but music with an excess.

As tempting as it may be to think of sound as the poetic excess of music, or of sound as a material thing that exceeds language, Lucier's early works and sketches explicitly swerved away from positioning sound as "real," instead valorizing its material, performative instantiation in the symbolic world. In an undated notebook page from this period, for example, Lucier articulated that the poetics of these works was "poetry derived from the person's relationship to the performance, not the relationships that exist among the sounds within the work."[103] The poetic excess present in *Music for Solo Performer* emerges not from the signals or even the sounds produced by the performer, but from both the performer's and the audience's socially and historically situated relationships to the performance itself. Since the sounds of these works seemed to matter less than their performance, Lucier experimented with poetic excesses such as amplified lips, profoundly dark squash games, and ring-modulated baseball broadcasts. Far from merely utilizing apolitical "sophisticated sensing devices" furnished to him through the military-industrial-academic complex to explore sound itself, Lucier seems to have been well aware of the technopoetics embedded in these devices' social, political, and historical contexts, which were thematized in the theatrical spectacles of his early works' performances in the 1960s.

In this sense, the instruments and instrumental systems employed by Lucier in the 1960s stray far from the composer's black boxes imagined by Subotnick, systematized by Buchla, or incorporated by Oliveros into her patchwork selves. The arc of this book began with the composer's black box as Subotnick's imaginary extension of the bounded human subject; this was contrasted with Buchla's expanded psychedelic subjectivity, unbound through a metaphorical expansion of its computational capabilities. Oliveros's "very unstable nonlinear music making system" expanded Buchla's exploration of the malleability of the human subject to the human body itself, developing patchwork subjectivities that incorporated various pieces of studio equipment. In all of these cases, the specific historical, social, and political contexts of instruments such as oscillators, tape delay, and other cutting-edge electronics were generalized into ostensibly universal questions about the human being as an abstract, monolithic object of scientific research. Lucier's diffractive work simultaneously broadened these questions even further to the nonhuman—dolphins, bats, and other "living creatures who inhabit dark places"—and also made them more specific to the historical context of the 1960s; each individual performer of his works developed their own poetics in relation to the work.

In the following chapter, I examine one final case of a composer's black box that similarly both broadened and specified the relationships between human beings and the technoscientific domains of information theory and cybernetics in the late 1960s and into the first years of the 1970s. As Oliveros conceptualized her work in relation to her body's gender and ability, and as Lucier conceptualized his work in relation to its performer's sociocultural and historical positionality, Sun Ra, the final composer studied in this book, also conceptualized his work in relation to the question of his subjectivity. But unlike Oliveros or Lucier, or Buchla or Subotnick for that matter, Ra had long been forced to think about his subjectivity through the ways it was socially constructed and racialized as Black, and the ways in which his subjectivity was differentially positioned within discourses of liberalism, universality, and even humanity in general.

In order to transcend the social and political failures that were so readily apparent to him as a Black subject on planet Earth in the mid-twentieth century, Ra understood himself to be a subject of the cosmos, beholden to a different kind of authority—articulated through electronic sound as a nonearthly, more-than-human force. While much of Ra's compositional work sought to thematize and illustrate that force through the tight control of his ensemble, the Arkestra, Ra's work with a prototype Minimoog synthesizer from 1970, contrastingly, opacified that sound, staging a refusal of legibility even to Ra himself in the moment of performance. Unlike the Arkestra, or even any of his other electronic keyboard instruments, the Minimoog refused to be controlled, developing, in Ra's words, "a mind of its own."[104] In this sense, for Ra, the opacity of his black-boxed, prototype Minimoog reflected a broader commitment to opacity in his socially pessimistic yet cosmologically utopian ideology. As was the case for Lucier, the sounds Ra produced with his black box were not "real," nor were they evidence of scientific claims about human consciousness, sensation, or perception—which for so long had relied on, and reinscribed, the racially unmarked liberal subject as a universalized model for humanity. Instead, Ra's black box produced both sonic and ideological opacities that pointed to sound and life well beyond those audible, or understandable, here on planet Earth.

5 Sun Ra and the Minimoog

FREEDOM, DISCIPLINE, AND OPACITY

"Sun Ra Coming!" So heralded the critic Richard Williams in *Melody Maker*, the English music magazine, on October 10, 1970.[1] Sun Ra, along with his twenty-one-piece Intergalactic Research Arkestra, had just begun his first European tour, nearly three decades into his career as a composer, bandleader, poet, and space-age philosopher.[2] Two days before the Arkestra's landing at London's Queen Elizabeth Hall, Williams provided some additional context to curious English music fans, most of whom likely had never heard Sun Ra's music, unavailable except through mail order—"the best kept secret in jazz."[3] After recounting a pocket history of Ra's career as a composer and bandleader in Chicago and New York and mentioning some of the space-age titles of his compositions, such as "Rocket Number 9" and "We Travel the Spaceways," Williams turned his attention to Ra's instruments in a section bearing the one-word heading "Moog": "As well as his roksichord [*sic*] and spacemaster [*sic*], Sun Ra will be bringing a Moog synthesizer to London with him."[4] The Moog synthesizer, still a relatively recent instrument, was having a moment; popularized around the globe by Wendy Carlos's 1968 hit album *Switched-On Bach* and further glamorized by English rock bands such as the Beatles in 1969, the sound of the Moog was quickly becoming characterized as the sound of the space age, even as the instrument was still relatively elusive. The arrival of Ra's Moog in London, to Williams, seemed to be just as important as the arrival of the man himself.

Indeed, Ra's Moog was mentioned in nearly every preview and review of Ra's London debut, becoming a metonym for Ra and the Arkestra's interstellar sound to English journalists.[5] In the *Guardian*, Ronald Atkins recounted of Ra's performance that "at one point, Ra himself improvised on his synthesizer; the noises were indeed not of this earth, but the intensity and cold logic of what he did chilled the spine."[6] In the *Times*, Williams similarly recalled that "a long, crashing solo on Moog synthesizer may be followed by a winsome ditty dealing with the delights of space travel, sung by the entire band, or a fearsome collective improvisation which tugs at the nerve-ends."[7] By the time of Ra's appearance in London, the dramaturgy of the Arkestra's theatrical stage show had been honed for nearly a month in other European cities, including Nanterre, Lyons, Paris, Donaueschingen, Berlin, Barcelona, and Amsterdam. This show, which included twenty American members of the Arkestra and two percussionists from West Africa, involved costumes, props, banners, processions throughout the audience, and a telescope aimed at Saturn.[8] Ra would lead the ensemble through moments of energetic group performance and choreographed big band numbers with call-and-response choruses featuring vocalists June Tyson, Gloristeena Knight (aka Ife Tayo), and Verta Smart-Grosvenor, alternating between apparent dynamics of chaos and control. Near the end of the set, after several verbal interstellar invocations by the Arkestra, Ra would play his Moog—always alone, in a solo that would range from two to twenty minutes in length, ending with an energetic "space chord" from the Arkestra as a grand finale.

Although Ra and the Arkestra's performances were chaotic, and his Moog solos especially so—both chaotic and unpredictable, even to him—at the beginning of his London performance on November 9, Ra reportedly proclaimed: "I demand discipline and precision!"[9] Ra's demand seemed to react to a number of elements of his tour that may have seemed undisciplined and imprecise—most immediately the misbehavior of many previous European audiences, who had variously rioted when venues oversold his concerts and just as often booed once they got inside.[10] But discipline and precision, a pair of concepts about which Ra had been publicly thinking on radio and in print since the mid-1960s, were also demands he made of his Arkestra, the ensemble that, to journalists like Williams, performed "fearsome collective improvisations." Ra insisted that these were not improvisations, but rather precise performances of his compositions—whether or not they were physically written—effected through the disciplined rehearsal of the Arkestra's members with their musical instruments. Indeed, in the program booklet for his London concerts, Ra explained, "I can write something so chaotic you would say you know it's not written. But the reason it's chaotic is because it's written to be. It's further out than anything they would be doing if they were just improvising."[11] Ra not only demanded discipline and precision from the Arkestra; he also demanded them of himself, especially in reaction to his Moog synthesizer—an instrument that was particularly imprecise and refused discipline because of its unpredictable internal electronic behavior.

Although Ra's Moog synthesizer looked a lot like his other electric keyboard instruments—which on this tour included a Farfisa combo organ, RMI Rock-Si-Chord (an electric harpsichord), Hohner Clavinet (an electric clavichord), Hohner Electra (an electric piano), and "Spacemaster" organ (likely a Gibson G-101 combo organ)—it was not an electric keyboard instrument, and it was also not exactly a Moog synthesizer.[12] That is to say, it was neither an electric organ, on which each key of the musical keyboard would reliably produce a note at its specified pitch, nor was it a large, modular electronic musical instrument with a keyboard controller whose individual modules could be patched together and precisely controlled to produce novel sounds, such as the Moog synthesizers heard on recordings by Wendy Carlos or the Beatles. It was, instead, a prototype of a new instrument designed not by Moog but by one of his engineers, an attempt to miniaturize, systematize, and integrate various functionalities of the larger Moog modular systems into a single box. Dubbed the "Minimoog" Model B (indicating that this was the second prototype), this instrument veiled the electronic processes by which it produced sounds instead of allowing its user to fully control the interconnections and parameter settings between modules—a literal example of *black boxing*, wherein a complex, nonlinear system is encapsulated and made opaque, reduced to input and output. Because the inputs into this instrument were imprecise, and because its circuits were constructed with errors that created nonlinear behaviors, its outputs were unpredictable, making it an apparently self-governing system with what Ra called "a mind of its own."[13] The musical keyboard and organ octave selectors present in its external user controls belied its idiosyncratic inner workings, which produced an output that could not be controlled in a disciplined or precise way.

In this sense, Ra's prototype Minimoog was, like Buchla's Box, Oliveros's patchwork system, and Lucier's experimental instrumental systems, a musical black box. Ra's relationship to this black box was, however, fundamentally different from the relationships between musicians and black boxes discussed in the other four chapters of this book. In those cases, musical black boxes were imagined in some way or another as liberating inner human subjectivity by allowing it to be more immediately expressed, prostheticizing and enhancing the modalities of musical expression available to the human subject, or otherwise freeing the subject from the confines of inherited models of musicality. Subotnick, Buchla, Oliveros, and Lucier expressed varying degrees of optimism about their black boxes, which they imagined could help them better understand, and potentially change, the fundamental structure of the self. For Ra, however, the opacity of the Minimoog seemed to help him articulate a radically different proposition about the self as unknowable, immeasurable, and potentially more-than-human or other-than-human. Instead of exploring the black box in order to understand its internal workings and develop plans to change it—turning it into what cybernetician Norbert Wiener called a "white box"—Ra embraced his black box's blackness, with no plans to turn it white.[14]

It is not a coincidence that the colors of these metaphorical boxes are the same as those used to demarcate racialized categories of identity in twentieth-century America. As Louis Chude-Sokei has shown, the legacies of histories of enslavement are omnipresent in canonic cybernetic texts from the twentieth century, not only in spectacular titles such as Norbert Wiener's 1950 book *The Human Use of Human Beings* but also in the fundamental metaphorical analogies that made cybernetics thinkable, such as master, slave, robot, and automaton. In all of these terms, Chude-Sokei writes, the "technological notion of otherness [...] can be traced back to the historical ground of the analogy itself: slavery."[15] Questioning "engineering colloquialisms" such as master and slave—or in our case, especially the colloquialisms of the unknowable black box and the knowable white box—can help us explore what Chude-Sokei calls the "intensity of metaphor in its ability to codify historical relationships"; such questions are "about the sedimented racism of seemingly lifeless objects, which because they are arranged in relationships with one another, cannot escape certain 'grammars' of dominance."[16]

For Ra, the opacity of the Minimoog as a black box was not an engineering challenge to be overcome, to be transformed into a white box. Instead, the prototype Minimoog's opacity was a fundamental feature, disallowing his performances from being collapsed into language, or even knowledge, and keeping them separate from extant ways of knowing and being in the world. Ra's refusal of earthly ways of knowing and being was a specific political reaction to the social conditions in which he lived in twentieth-century America; while musicians such as Subotnick, Buchla, Oliveros, and Lucier could optimistically imagine social futures for their apparently electronically liberated subjectivities, and indeed did enjoy celebrations of their musical and technical liberations in American and international musical institutions throughout the twentieth century, Ra was much more pessimistic (even as, after his death, he would become lionized as a patriarch of Afrofuturism within American cultural institutions). Freedom, for Ra, was an illusion that could lead only to social death on Earth; true emancipation could only come through a disciplined and precise adherence to a higher cosmic authority.

In 1971, after a year of using his Minimoog, Ra explained his positions on freedom, discipline, and precision in a live radio interview:

> Now they've been read up on the word freedom, which is what the white people have been selling. Now when they deal with freedom, they can't play my music, because I don't care anything about freedom. My whole thing is based on discipline and precision, because that's what nature does. There's no freedom in nature. You have your particular place, and you do what you're supposed to do. As far as I'm concerned, the only freedom that I've ever seen that Black people get has been over in the cemetery, and the only peace I've ever seen them get is that "rest in peace" that the man reserved for them.[17]

Ra had been publicly discussing discipline and precision as the key to transcending earthly freedom and death as early as 1967. Journalists such as the Nigerian writer Tam Fiofori had understood the vessel for such discipline and freedom to be the Arkestra, which by using its instruments "as a straight-ahead channel/medium," and by "relating to Run Ra intuitively [...] was initiated as a unit to master the principles of Sound—its essence as the source of music and magic."[18] With the Minimoog, however, Ra encountered an instrument that could not be disciplined and refused to be precise. Ra's performances with this instrument, as I show in this chapter, articulate Ra's philosophy of discipline and precision in real time—not only in their sound, but also in their dynamics of control.

To understand how Ra's prototype Minimoog became a black-boxed articulation of his concepts of discipline and precision through its opacity, this chapter first contextualizes Ra's encounter with instruments designed by Robert Moog by exploring Ra's long-standing interest in new electrical and electronic instruments throughout the 1950s and 1960s, exploring his first encounter with a Moog modular synthesizer in 1969 before his later visit to Moog's factory in Trumansburg, New York, in 1970, where he acquired his prototype Minimoog. It then shows how Ra utilized the Minimoog in a cybernetic, black-boxed modality of live performance throughout 1970 and 1971, specifically through a close reading of Ra's first public performance with the Minimoog in August 1970. The chapter analyzes Ra's performances with the Minimoog in this period through his concept of discipline, showing how it relates to the opacity of the black-boxed Minimoog. By following connections between Ra's cosmology and the metaphorical and conceptual world of cybernetics, it ultimately shows how Ra's use of the Minimoog produced a cybernetic subjectivity that was opaque, swerving from optimistic possibilities for life on Earth and instead embracing what he called an "alter-life" rooted in "myth-science," an alternative path forward for the Arkestra and like-minded fellow travelers.

RA'S EARLY RELATIONSHIP TO TECHNOLOGY

Over the course of his nearly sixty-year career, Ra consistently dedicated himself to exploring new electrical and electronic musical technologies, often acquiring new keyboard instruments the moment they became commercially available. After his death in 1993, scholars and critics have pointed to this dedication to celebrate him as a "pioneer in using various—then weird—electronic instruments," as Amiri Baraka wrote in 1995.[19] Since most of these instruments were keyboard instruments, and since Ra was a pianist and organist, they have often been celebrated for their "multiplicity of sounds" rather than for their specificity as individual instruments.[20] Indeed, their novel timbres allowed many listeners to easily associate Ra's music with other uses of electronic sound in the 1950s and 1960s—namely "evoking the alien," as Kodwo Eshun put it in 1998.[21] In this reading, the Minimoog was yet another of a string of novel electronic instruments that Ra used to explore new extraterrestrial worlds of sound.

But the Minimoog was different in kind, not degree, from Ra's other instruments. Those instruments, including the Solovox, the Clavioline, the Gibson/Kalamazoo 101 electric organ, and the Rock-Si-Chord, were designed as relatively traditional keyboard instruments. When a key on their musical keyboards was depressed, a mechanism was activated that produced a sound that matched the pitch of the corresponding key or keys. Although some were monophonic, meaning that only one key could be active at a time, all were designed to produce "musical," that is, harmonic or pitched, sound; with polyphonic instruments such as electric pianos or organs, Ra could essentially use the instrument just as he would an acoustic piano or organ, providing harmonic accompaniment and performing solos in his compositions with the Arkestra.

Although the Minimoog featured a musical keyboard, the sounds produced when its keys were depressed were determined by a matrix of parametric knobs and switches that allowed the user to produce an extraordinary range of sounds—from pitched to unpitched, clear to noisy, stable to warbly. Ra could not play chords on the Minimoog; its monophonic keyboard and imprecise, knob-controlled parametric settings meant that even playing a melody in tune with another instrument was extremely difficult. Because of this, between 1970 and 1971 Ra almost exclusively performed with the Minimoog as a solo instrument, without the Arkestra. His performances with the instrument, as I document in this chapter, are unique articulations of his concept of discipline, producing a cybernetic subjectivity that valorized its own opacity—or, as he put it in his 1974 film *Space Is the Place*, the "intergalactic realm of eternal Black darkness."[22]

This opaque, cybernetic subjectivity could not specifically be produced with the electric instruments that Ra played before the Minimoog. Instead, throughout the first two decades of his career, Ra used electric instruments in his compositions to produce timbral and visual representations of technological futurity that could be readily understood by his audiences and fully controlled by Ra himself. Even as early as 1942, as a bandleader in Birmingham, Alabama, Ra supplemented the sound of his piano with the Solovox, an early monophonic three-octave electronic keyboard instrument that sat above the keyboard of his acoustic piano. In the words of band member Frank "Doc" Adams, the Solovox "made organ-type sounds"; it was employed not only for its timbre but also for its visual novelty.[23] One advertisement for an evening's entertainment at Birmingham's Congo Club in July 1942 lists Ra's ensemble as Sonny Blount's Solo Vox Band as a part of a "variety show of Live Wire Entertainment," suggesting that the instrument was a selling point in the "wired," technological spectacle of Ra's live performances.[24]

Later that decade and into the 1950s, Ra continued to use electronic instruments to evoke technological futurity through instrumental timbre. Ra took the Solovox to Chicago, where he began to make recordings in his own apartment with an early wire recorder.[25] One of the earliest such recordings from 1948 was a duet with violinist Stuff Smith of Peter DeRose's tune "Deep Purple."[26] The recording, eventually released on Ra's Saturn Records label in 1973, is a relatively straight-ahead rendition of this popular song; Ra treads the Solovox as a melodic instrument, first playing the tune's melody with an organ-like timbre with vibrato, accompanied by his own left hand playing chords at the piano and Smith's violin pizzicato. Although the timbre of the Solovox in this early recording was novel, Ra was essentially using it as a stand-in for an electric organ, just as he had in Birmingham.

As Ra's live performances became more radical in the mid-1960s, so did his use of new electric instruments. By that time, the American musical instrument market had become flooded with new keyboard instruments, fueled by the sales of European patents to American musical instrument companies and the novelty of records such as Joe Meek's 1962 hit single "Telstar," which featured a new instrument called the Clavioline.[27] Ra, of course, purchased one as soon as it was available. The Clavioline was another keyboard attachment instrument, related to the Solovox, that was intended to supplement a pianist or organist's timbral palette, with a limited-range musical keyboard and a knee controller for volume. Unlike the Solovox, however, the Clavioline had vastly expanded controls for timbre, pitch, and vibrato, with some models featuring up to twenty-two timbre switches that functioned as organ stops. The Clavioline's timbre was piercing and shrill; its vibrato was significantly deeper than other instruments, exaggerated in the manner that had become popular on instruments such as the theremin.

Ra explored the timbral possibilities of the Clavioline on the A side of *The Magic City*, his landmark album released in 1966. This side contains 27.5 minutes of audio recorded at Babatunde Olatunji's Center of African Culture in Harlem, wherein the Arkestra performs what sounds like a free improvisation, although *The Magic City* as a composition was completely scored, its performance tightly controlled by Ra with hand signals, cues, and other gestures.[28] Throughout *The Magic City*, the harsh, throaty, warbly sound of Ra's Clavioline dialogues with his own piano playing, as well as with the flutes, saxophones, trombones, and bowed bass of the Arkestra, all fed through layers of reverberation and tape delay. The extreme depth of the Clavioline's vibrato bears a striking resemblance to that of the theremin, leading one critic to describe it as the sound of a "horror movie."[29]

In May 1967 Ra began to develop another work at the Olatunji center featuring his new Gibson/Kalamazoo 101 electric organ, which he dubbed the Solar Sound Organ.[30] This work, entitled *Atlantis*, uses the organ as a mimetic sonic representation of the lost continent of Atlantis, with Ra maximizing the use of the instrument's organ stops to create an additive flood of sound. A recording of *Atlantis* made during its August 1967 premiere begins with a short, chaotic drum solo but quickly narrows down to Ra, solo, on the Solar Sound Organ. In its opening section, Ra plays a repeated, single A, but each time the note is repeated, Ra pulls out another stop on the organ, so that the apparent pitch is first doubled at its octave (the 8′ stop), then its fifth (the 2 2/3′ stop), and so on up the harmonic series.[31] As Ra's biographer John Szwed reads the work, it progresses from single pitches to a "virtual sonic representation of the flooding of Atlantis"; after Ra has seemingly pulled out all the stops, he further intensifies his performance by using extreme, whole-hand and -arm gestures on the keyboard in a frenzied solo.[32]

Szwed likens *Atlantis* to J. S. Bach's Toccata and Fugue in D Minor (BWV 565), and the comparison is apt. Not only do both compositions express raw sonic power, but they also explore the timbral limits of the instruments at hand. Ra's Solar Sound Organ allowed its player to choose from three voices for its bass keys and twelve voices for the treble keys, plus effects such as vibrato and sustain with discrete settings; throughout the composition, Ra seems to maximize the combinatorial possibilities of the instrument's timbre. Even as this timbral exploration is taken to mythical, liminal spaces in *Atlantis*, though, the Solar Sound Organ was still limited by the fact that its controls were only controllable in discrete steps determined by the instrument's designers. And in the same period that Ra was creating his Bach-like *Atlantis*, another composer in New York City was using a new kind of electronic instrument to bring her own realizations of Bach compositions into the world. That composer, Wendy Carlos, and the album she produced in early 1968, *Switched-On Bach*, would introduce a new category of electronic instrument to most of the world: the synthesizer.

MEETING THE MOOG

Carlos's "switched-on" realizations of Bach compositions were created using a system of interconnected modular electronic instruments designed in collaboration with Robert Moog, whose name was quickly becoming a generic noun for his instrument. Carlos met Moog at the fall 1964 convention of the Audio Engineering Society, where Moog had somewhat accidentally found himself an exhibitor, offering several voltage-controlled modular instruments—similar, technically, to the voltage-controlled modular instruments designed by Donald Buchla in 1963–1964, discussed earlier in this book, although with orthogonally different imagined use cases.[33] By early 1966, Moog and Carlos were collaboratively designing a customized system for Carlos's personal use; by 1967, it was installed in her apartment on the Upper West Side of New York City, and throughout the following year it was used to create *Switched-On Bach*. Released in October 1968 and certified as a gold record by August 1969, the album quickly launched the Moog synthesizer into national and international media discourse.

Yet even as Moog was becoming a household name, it was unclear to most people how his new instrument worked. Unlike an electric organ such as the Gibson 101, Moog conceptualized his synthesizers as "integrated systems of modular, single-function generating, modifying, and controlling instruments."[34] Conceptualized more like general-purpose computers than keyboard instruments, Moog wrote in 1967 that "all of the instruments comprising the Moog Synthesizers are designed for as general application as possible."[35] Although Moog specified that "the Synthesizer should be readily adaptable to both live performance and programmed control," during the first few years of production, Moog synthesizers were nearly exclusively used in the latter application. Installed in recording studios in New York City and Los Angeles, they were programmed by specialists, their sounds recorded onto tape for use in the construction of multitrack recordings. Indeed, the first eight commercially released records featuring Moog synthesizers, all produced in 1967, were programmed by Paul Beaver, an organist, session musician, and Scientologist who had become Moog's West Coast sales manager.[36] Carlos's *Switched-On Bach*, produced in 1968, was created using a painstaking process of programming the timbre of each individual voice from Bach's polyphonic works on her Moog system, playing through each line monophonically and then carefully overdubbing and bouncing each individual voice onto multitrack tape.

Media representations of the Moog synthesizer almost always fundamentally misconstrued the way it operated, representing it as a new musical instrument akin to the saxophone or the violin.[37] Throughout 1968 and 1969, articles about Moog's synthesizers proliferated in outlets ranging from newspapers such as the *New York Times* and the *Saturday Review* to trade periodicals such as *Electronics Magazine* and *System*, the in-house magazine of Consolidated Edison.[38] By April 1969, the Moog made it to New York City's Lincoln Center for a Young Person's Concert broadcast live on CBS; Leonard Bernstein joked about how this new instrument could "do anything but stand up and take a bow."[39] Although Bernstein played the audience a prerecorded tape of a realization of Bach's "Little Fugue" in G minor (BWV 578) from a small tape machine perched on top of the giant Moog console, he also switched on the Moog's sequencer module, creating an appealing cascade of blinking lights that allowed his audience to believe that the instrument was in some sense playing itself.

It was in this moment of national attention, speculation, and fantasy about the Moog synthesizer that Ra sought out an encounter with this new instrument. Through the engineer Fred Vargas, who recorded and often housed the Arkestra when they were in New York City, Ra was introduced to the composer Gershon Kingsley, who had been the first musician after Paul Beaver to program his own Moog synthesizer for performance on a record.[40] On November 12, 1969, Ra arrived at Kingsley's studio with John Gilmore, Marshall Allen, and Danny Davis of the Arkestra and recorded four takes with Kingsley's Moog system. But although this session marked Ra's first encounter with a Moog synthesizer, it is likely that Ra did not interact with the instrument beyond playing its keyboard(s). As had been the protocol for the brief few years of the Moog's existence, the Moog's parametrical controls and settings were programmed exclusively by Kingsley, and Ra simply played it like a keyboard instrument. With his hands on two monophonic keyboards controlling different banks of oscillators, Ra created an incrementally spaced-out version of his familiar musical style. These timbres remained static throughout all four final takes produced during the session, which suggested that the Moog's controls were not touched during Ra's performance. In several outtake reels discovered by Ra's estate, Kingsley's voice is heard in the studio, suggesting that he stuck around to monitor Ra's use of the instrument and to program it between takes.[41]

It is tempting to imagine Kinglsey bonding with Ra over their shared interest in new, electronic musical timbres. But in a 2016 conversation with Irwin Chusid, who began managing Ra's estate in 2013, Kingsley remembered the session with a relatively dismissive tone:

> One day Sun Ra—or one of his associates—approached me and said, "we would like to record in your studio." I said "go ahead." [. . .] I don't know who mixed it all together, and don't remember if they paid me or not. I only met Sun Ra during his sessions to preset the Moog Modular for him. It wasn't an important session to me, and I didn't spend much time on setting it up. He came in, said he wanted to play the Moog and record with it. I showed him what to do. I don't know if you'd call that a "collaboration" or a "duet." I helped him, but I didn't do anything with his music.[42]

Kingsley's lack of enthusiasm about Ra—a person one would imagine it would be difficult to mistake for "one of his associates"—hints at the state of alterity that Ra would soon find himself inhabiting within the social world of Moog synthesizers. Unlike a Solovox or Clavioline, a complete Moog system could not be purchased in a store and could cost in the tens of thousands of dollars; as of January 1969, fewer than ninety individual musicians owned Moog systems worldwide (and quite likely even fewer than that, since this figure, estimated by Moog, also included owners of commercial music studios who were not necessarily musicians themselves).[43] Among this "costly cathode club," as the journalist Harvey Siders characterized it in 1970, there were only three jazz musicians: saxophonists Oliver Nelson and Benny Golson, who had each purchased complete systems and hoped to make up their investments with commercial studio work; and the pianist Paul Bley, who claims that he was given a complete Moog system for free as a "road tester" in 1969 directly by Moog himself.[44]

Whether the prices that Nelson, Golson, and Bley paid for their Moogs were influenced by their identities—Bley being a white Canadian and Nelson and Golson being African Americans—is impossible to address, based on the scant historical record of any of these interactions. But Moog's encounter with Ra in the summer of 1970, mere months after he apparently gave away a full modular system to Bley, seems to suggest that Moog's attitude toward what kind of musicians ought to own his instruments—and, as we will see, *which* instruments they should own, and for what cost—was at least partially influenced by his perception of their identity.

In early 1970, a few months after Ra's session with Kingsley, the young Nigerian writer Tam Fiofori, who had been working for Ra as a booking agent, developed a plan to broker a meeting between Ra and Moog. Fiofori had been preparing a special issue on electronic music for *DownBeat* magazine, and under the magazine's aegis, he flew to Trumansburg, New York, to interview Moog and convince him to give a system to Ra. As Fiofori recalled in 2004, although his trip was ultimately successful, it was marked by Moog's tacit discomfort in dealing with Fiofori and the Arkestra—largely, in his memory, because of their racial identities. After being "the only Black face on the flight" from New York City to Ithaca in the summer of 1970, Fiofori waited as all of the other passengers were met and picked up; when the "one worried chap who was pacing up and down"—Moog—finally worked up the courage to approach him, Fiofori could see his "face, and I'm sure his heart, drop."[45] To Fiofori, it quickly became apparent that Moog had assumed his name was Italian, not Nigerian. After an uneasy meal of pasta and red wine at Moog's house, Fiofori remembered Moog complaining to him: "I am still waiting for that musician who has the sound-scope to play the synthesizer as a musical instrument, not an effects box." Fiofori responded: "'I know how I can solve your problem,' I assured him. 'Who?' He asked eagerly. 'Sun Ra,' I told him. 'Sun Who?', he retorted." Once this introduction was made, Fiofori called Ra, who drove up from Philadelphia with several members of the Arkestra the next day.[46]

Moog had been keeping meticulous records of potential customers, calls, inquiries, and correspondence throughout this period, yet in the 263 linear feet of paper in Moog's archive, there is no trace of Ra's visit to Trumansburg.[47] Indeed, the first time Moog publicly discussed Ra's visit was in 1991, in a brief interview with Sun Ra superfan John Hinds for his self-published *Sun Ra Research* periodical. Moog remembered: "[Ra's] whole band came to this small upstate town in New York. It was, I don't know, fourteen or fifteen people, and they came up in three or four cars—each one older than the other. [...] It was a piece of ... sociology—a demonstration of a piece of culture that is never seen in upstate New York."[48] Moog's euphemistic language describing Ra and the Arkestra's visit as "a piece of sociology" suggests that in his memory, Ra was not exactly welcome in Trumansburg. This is corroborated by the memories of Moog's employees; engineer Jon Weiss, for example, remembered that "[Ra] came to Trumansburg where the Moog company was back in '68 [*sic*], and this was a fairly rigid, straight-laced, sleepy little New York State town. And here's this bizarre looking black guy with, you know, robes and all this stuff in the local ice cream parlor!"[49]

Apparently, when Ra and the Arkestra arrived in Trumansburg on July 29, 1970, Moog had them wait in an ice cream parlor before allowing them into his factory; the presence of the Arkestra produced many "colorful" anecdotes in the memories of Trumansburg residents that articulated overtly anti-Black racial stereotypes. As Brian Kehew, who has served as an official historian for the Bob Moog Foundation, wrote, "The town's phone lines lit up with news of the visit. Shortly after, reported bounced checks from the group's visit added to the consternation, and one local knitter claimed a sweater was allegedly taken that was never paid for."[50] When Ra did make it into Moog's factory, he was apparently not shown the flagship Moog modular synthesizer. Instead, he was shown a prototype of a Miniature Moog, or Minimoog for short, an instrument that one of Moog's engineers had been working on in his spare time, kludged together from parts from Moog's "synthesizer graveyard" and not yet fully operational.

Figure 19. The Minimoog Model B, 1970. Image courtesy of EMEAPP.org.

Like Ra, the Minimoog also inhabited a state of alterity in Trumansburg in the summer of 1970. Though it would eventually become one of the most iconic keyboard synthesizers of that decade, at the time of Ra's visit the instrument was still a prototype. Bill Hemsath, a recently hired engineer, had begun combining old, discarded synthesizer parts into an integrated instrument that did not use separate modules, but instead brought together oscillators, filters, envelope generators, amplifiers, and a keyboard into a single housing. Assisted by another recent hire, Jim Scott, Hemsath built what he called the Model A Min in November 1969 and had continued to develop the instrument with a second prototype, the Model B (see figure 19), by the summer of 1970. (By this point, the instrument came to be called the Minimoog, rather than the "Min" Moog.)[51] Robert Moog, who was not involved with the design of either of these prototypes, remembered the project dismissively: "Hemsath was always doing shit like that. [...] I didn't have too much time to think about doing stuff like that."[52] Indeed, in 1970 Moog was focused on designing larger-scale systems such as an integrated tape recorder and studio mixing console specifically designed to work with his synthesizers, as well as a computerized synthesizer.[53] These projects, however, were financially unsuccessful, and a bank loan application saved in Moog's archive suggests that the company needed a massive capital investment to remain solvent.[54] Faced with this dire financial situation, Moog saw Hemsath's Minimoog project as a distraction from larger, higher-priced systems that would generate more revenue, but Hemsath, along with the other engineers, continued to clandestinely work on the project when Moog was not present.[55]

Why was Ra shown this prototype instead of the flagship Moog modular synthesizer? Most accounts of Ra's time in Trumansburg gloss over this question, simply celebrating the fact that Ra seemed to be taken with the Minimoog prototype and that Moog seemed to be happy with Ra's use of the instrument. Tam Fiofori, who brokered this encounter, remembered in 2004 that Moog was "speechless, stopping short of hugging Sun Ra," and that he "spontaneously offered his [sic] first [sic] prototype Minimoog completely FREE to Sun Ra."[56] For his part, Moog downplayed the encounter and focused on the financial dimension of this transaction, remembering that "we loaned it to him, and Sun Ra's way of working is that when you loan something to him you don't expect to see it back (laughs)."[57] While there is no evidence to suggest that Moog held any inherent bias against Ra, the fact remains that Ra left Moog's studio that day with a barely working prototype of an instrument that Moog thought was robbing his engineers' time from his larger systems. Both Ra and the Minimoog held a status of alterity, both musically and technically; it wasn't exactly clear what they were doing, or how they worked.

Archival audiotapes recorded during Ra's visit to the Trumansburg studio evince this alterity, bearing the initial traces of a radically new modality of performance with the Minimoog for Sun Ra. Unlike Ra's performances with the larger Moog modular, in which he used the instrument's keyboard to produce harmonic sounds with novel timbres, Ra's first performances with the Minimoog produced inharmonic, noisy sounds with constantly changing timbres. This is likely because the instrument's controls were ready to hand, located directly above the keyboard, unlike the Moog modular system, in which the keyboard was separated from the individual modules' control. With the Minimoog, Ra could easily manipulate the keyboard and the controls at the same time, which he did. While these recordings were most likely made for archival purposes and never intended for public consumption, Ra (or his business associate Alton Abraham) did choose to include them on a later commercial release in 1971 entitled *My Brother the Wind Volume II*, suggesting a sequel to the 1970 album he had previously recorded in Kingsley's studio in 1969, *My Brother the Wind*.[58] Although this editorial choice suggests a continuity between Ra's use of the Moog modular and the Minimoog, as I will show, it was only with the Minimoog that Ra began to embark on a fundamentally new way of performing music.

Unlike the larger Moog modular's ability to be precisely configured and reliably controlled, the prototype Minimoog was more akin to a black box. Kludged together from spare parts, its oscillators did not track properly, its filter section was erroneously designed with 15 dB of extra gain, and its resurrected components added to its unique complexity and nonlinear behavior. It was precisely this alterity that made the prototype Minimoog an essential part of Ra's performance practice in the months and years to come. In 1971, after a year of playing this instrument, Ra boasted:

> I got one like nobody else has in the world. [...] [T]his is one that ain't nobody has one, like I say I don't want one like anybody else got, because they just threw it together for me when I went up there one Sunday, just run and put it together, and actually they want it back, because actually they don't know what it does, really.... [T]hey ask me and I say it does lightning and thunder. I hit a note here, and three minutes later it starts lightening and thundering on them, and I don't.... It just does it, like it's got a mind of its own.[59]

From August 1970 through the autumn of 1971, Ra would use this instrument in a way unlike any of his previous electric keyboards, reserving it for solo performances of new music that bore no relation to his extant compositions. In his emergent performances with the Minimoog, Ra produced sounds outside of the equally tempered scale, and indeed outside of the realm of harmonicity. To Ra, the instrument seemed to "have a mind of its own," following its own self-regulating, cybernetic logic. Yet even as Ra began to negotiate his own compositional agency between himself and the Minimoog, he began to conceptualize his performances with the instrument as examples of "discipline," an adherence to a cosmological "nature" in which every subject "has their particular place" and "does what they're supposed to do."[60]

BECOMING THE BLACK BOX: *NUITS DE LA FONDATION MAEGHT*

Ra's first public performance with the Minimoog took place the evening of August 3, 1970, five days after he had acquired the instrument in Trumansburg. Ra and the Arkestra had been invited to the Fondation Maeght, a recently founded contemporary arts foundation in Saint-Paul de Vence in southeast France, to perform a series of concerts as a prelude to their European tour that would take place later that fall, discussed earlier in this chapter. For the Maeght performances and that tour, Ra's new Minimoog was prominently featured in publicity photographs, concert previews, and eventually concert reviews (see figure 20).[61] Indeed, John Szwed's narrativization of the Maeght performances placed the Minimoog in the metaphorical and physical center of the Arkestra: "Nineteen musicians and dancers were on the stage on the evening of August 3. The audience had little or no knowledge of Sun Ra's music, since his records weren't widely distributed in France, and when they arrived they saw the Arkestra spread out before them like elaborate decor: musicians in red tunics, seated in a forest of instruments on stage, dancers in red dresses [. . .] and in the still center of it all, Sun Ra sat behind the Moog, creating the sounds of gales, storms, and waves crashing."[62]

Swed's narrative suggests that the Moog acted as a centralized instrument of control for the world of sound inhabited by the Arkestra during the performance. Yet listening to audio recordings of these performances reveals that Ra did not use the Minimoog to play his compositions with the Arkestra at all. Instead, he almost exclusively used it as a solo instrument for playing entirely new music, without precedent in his extensive compositional oeuvre.[63] With the Arkestra, Ra performed in his more familiar, almost Ellingtonian mode, with the entire ensemble performing Ra's compositions replete with vamps, sectional riffs, and repeating vocal phrases—and occasional excursions into freer group performance. To perform these

Figure 20. Sun Ra and the Minimoog at the Fondation Maeght on August 3, 1970. The Minimoog can be seen placed on top of an RMI Rock-Si-Chord; an electric organ, probably a Farfisa, is in the foreground. Photo by Jacques Robert, Archives Fondation Maeght, Saint-Paul-de-Vence (France).

compositions, Ra used his electric keyboards to play harmonic accompaniment, melodically and harmonically complex solos, and also his more chaotic, gestural solos; as in his 1967 composition *Atlantis*, Ra is heard playing cluster chords with his fingers, hands, and forearms, and building up his electric organ's timbre by adding additional pitches and stops to accompany his composition as played by the Arkestra. But with the Minimoog, Ra played alone.

During this first concert at the Fondation Maeght, and well through 1971, Ra developed a distinct performance practice with the Minimoog, exploring inharmonic sound without the Arkestra. Because of the Minimoog's parametrical design for control, its monophonic keyboard, and its easy slippage into inharmonicity, it could not easily perform any of Ra's previously composed works. Practically speaking, it could not play chords; it could not reliably play melodies in tune; and its 25-key monophonic keyboard meant that Ra could not use it like his other electric keyboard instruments. What it could do, however, was produce an emergent, cybernetic performance, outputting sounds that Ra could not have anticipated. Those sonic outputs would then steer Ra in an exploratory mode, suggesting that he play with the instrument's knobs and control dials in a reiterative, cybernetic modality. By allowing the sound that the Minimoog produced to influence the direction of his performance, Ra entered into a systematic relationship with the Minimoog in which his own actions, the electrical waves those actions produced, and the sounds that those electrical waves eventually became all functioned as informational signals that fed back between Ra and his instrument. In other words, they became a cybernetic, or self-governing, system, with the Minimoog behaving like a black box—an opaque system that was perceptible while remaining unknowable.

Although it is possible that Ra used the Minimoog to perform other compositions during the Maeght residency that were simply not recorded, on the two commercially available live recordings of the three evenings of that residency, there is only one track where the Minimoog is played, entitled "The Cosmic Explorer" on *Nuits de la Fondation Maeght*, Vol. 1. This is essentially a solo performance between Ra and the Minimoog, with brief, fleeting, gestural flourishes from other members of the Arkestra; at nearly twenty minutes, it offers a prime example of the emergent, cybernetic performance practice that Ra would continue to develop with the Minimoog throughout the following year on his European tour.

Ra begins his solo on "The Cosmic Explorer" by exploring the low end of the Minimoog's frequency range, mixing together the signals of its oscillators and its white noise generator. In real time, Ra depresses a single key on the keyboard to play one sound and simultaneously turns knobs to change the filter frequency, resonance, and amount of noise; he also quickly begins to turn the knobs of the frequency controls for the oscillators, resulting in sounds that glide between pitches. Since the Minimoog's keyboard was monophonic, meaning that the frequency of its oscillators could only be controlled by one key at a time, Ra could not play this instrument like any other electric keyboard. Indeed, very quickly into this performance,

Ra seems to embrace this idiosyncrasy, embarking on a rapid-fire passage wherein the instrument's pitch jumps around at an inhuman rate as the keyboard struggles to register his rapidly moving fingers. We hear Ra briefly developing a motivic melodic theme before again experimenting with the idiosyncrasy of the keyboard, with a mixture of two oscillators set to different frequencies, before he begins to experiment more intently with adding in noise.

During these noisier moments of inharmonic sound, one or two members of the Arkestra briefly add in a cymbal swell or a single high saxophone note, but these additions quickly fall silent as Ra continues with his solo voyage. After five minutes, the Minimoog's inharmonic noise seems to have caught Ra's attention. He begins to embrace the Minimoog's extremely noisy timbres—likely a result of the prototype instrument's faulty schematics or idiosyncratic construction.[64] With the instrument's filter resonance set to its maximum value, Ra sweeps the filter's frequencies to its upper and lower extremes and maximizes the addition of white noise; he sets the keyboard to glide between frequency values at slower and faster rates, creating meandering glissandi whose rate of change between pitches varies wildly. After about ten minutes of performing with the Minimoog, Ra adds in gestural whole-arm soloing on the Farfisa organ and Rock-Si-Chord keyboards with his other hand, continuing to perform occasional filter-swept glissandi from his Minimoog. And after seventeen minutes of solo performance with his Minimoog, Ra allows the Arkestra to join in at full intensity, with a nearly two-minute explosion of maximally intense performance before a sudden, synchronized end.

Recordings of performances from the Arkestra's fall 1970 tour suggest that Ra continued this emergent cybernetic performance practice with the Minimoog almost entirely alone. In recordings of sets from the Donaueschingen Festival, the Berlin Jazz Festival, and London's Queen Elizabeth Hall, ranging from late October to early November 1970, respectively, the Minimoog sits silent until the final quarter of the Arkestra's set, when it is used for an extended solo with no predetermined compositional material, with the Arkestra occasionally joining in for the performance's final moments.[65] As the tour progressed, Ra would increasingly explore the Minimoog's capabilities of producing inharmonic sounds by experimenting with combinations of its oscillators' frequencies, their waveshapes, the instrument's noise generator, and its filter. By November 9, during his performance in London, Ra used the Minimoog almost exclusively to produce harsh inharmonic sounds, with the keyboard's portamento controls set to make the oscillators glide between frequencies at extremely slow speeds.

Ra's performance practice with the Minimoog, like his previous performance practice as the leader of the Arkestra, was emergent and chaotic, yet also apparently self-governing through its cybernetic discipline. But Ra's Minimoog, because of its idiosyncratic design, did not behave like any other instrument—acoustic, mechanical, electric, or human—that Ra had previously used to perform his music. With the Arkestra, Ra functioned as the architect of a cybernetic system of individual players, highly trained to develop a sense of collective consciousness and emergent performativity. As Daniel Kreiss has interpreted, Ra "functioned as a designer who convened a system within which an emergent relationship between individuals was created through engagement with the technologies of musical production."[66] If the "people [were] the instrument," as Ra wrote in 1972, then the Arkestra functioned like a system of interconnected instruments under his architectural control.[67] The Minimoog challenged this notion of architectural control, leaving Ra to come up with a cosmological explanation. When he played with the Minimoog, Ra submitted himself to a higher cosmological authority, allowing himself to channel "nature" in his performances.[68]

Both Ra's control and the control of nature produced complex performances that seemed to be chaotic and irreducible to knowledge. However, Ra maintained a claim to authority over these performances. As he wrote in a program note for his November 9 performance at Queen Elizabeth Hall, "I can write something so chaotic you would say you know it's not written. But the reason it's chaotic is because it's written to be."[69] Ra's insistence that everything performed by the Arkestra was "written to be" conflates his own authority over the compositions performed by the Arkestra with a higher order of authority from the cosmos. In a similar way, Ra also claimed that his performances with the Minimoog were also channeling such a cosmological order. Both claims leverage the idea that these performances were "precise" and "disciplined" over the appearance of improvisatory freedom.

In a 1972 interview with Tam Fiofori, Ra explained: "With electronics [*sic*] instruments I can do what I call extensive continuation and infinity.

I'm demonstrating infinity, because that's the way the Universe is, and it keeps demonstrating that . . . the precision and discipline of it."[70] Between Ra's acquisition of the Minimoog in July 1970 and this interview, Ra increasingly began to contrast "precision and discipline" to what he considered the falsity of freedom. Indeed, in the same interview with Fiofori, Ra opened the conversation by insisting that "most people speak of freedom, but really the foundation of all freedom is discipline. The universe itself demonstrates a discipline and nature demonstrates discipline everywhere you look. Nature variates things, and it has this precision and discipline in doing this."[71]

Why did Ra increasingly valorize discipline and precision over freedom in the period immediately following his encounter with the Minimoog? In addition to responding to (and disavowing) the increasingly popular term *free jazz*, which by 1970 had become a catchall that lumped together widely disparate musicians and practices into an easily digestible commercial genre category, I suggest that the Minimoog presented Ra with an ontological challenge that forced him to more strongly articulate his positions on freedom, discipline, and precision. Unlike his other electric keyboard instruments, and unlike the members of the Arkestra, Ra could not precisely control the Minimoog in an observable way, and therefore it challenged his claims to compositional authority. Because of the instrument's instability and nonlinearity, he could not compose music for it that could be precisely performed, as he could with the Arkestra. Instead, the Minimoog seemed to have "a mind of its own," under a cosmological control that could not be reduced to knowledge or representation. To insist on the "discipline and precision" of performing with such an unstable instrument was to characterize Ra's own actions as being under a similar cosmological control, "demonstrating infinity [. . .] the way the Universe is."[72] The perceptible chaos of a performance with such an instrument masked what Ra insisted was an internal "precision and discipline" shared between him and his instruments, together in a systematic, self-regulating, emergent performance.

In the period immediately after acquiring the Minimoog, Ra began to increasingly intensify his valorization of cosmological discipline over the appearance of earthly freedom. In the closing section of this chapter, I show how the Minimoog provided a new model for human alterity—a condition that Ra had already been thinking about for years before his encounter with the instrument. The black-boxed nature of the Minimoog led Ra to conceptualize a new subjectivity that itself could be thought of as a black box—an opaque instrumental system that swerved from claims to humanity by refusing to be reduced to existing systems of knowledge or being.

PHYSICAL AND METAPHYSICAL DISCIPLINE

The word *discipline* held both physical and metaphysical meanings for Ra in the period immediately after encountering the Minimoog. Since the 1940s, Ra had been following an applied ethics of disciplined asceticism in his personal life that he also demanded of the Arkestra, which involved abstaining from intoxicants, sex, and meat, and (possibly apocryphally) rehearsing for up to twelve hours a day. According to Szwed, even in the earliest years of his career, he "lectured [the Arkestra] on personal discipline," insisting that that they be "vegetarian and eat natural foods." In Szwed's estimation, Ra maintained the discipline of the Arkestra so that "they would be an example to the world of what a group of black men could achieve together, an echo of what the great dance bands had achieved under strict, intelligent leadership."[73] From the 1950s through the 1970s, Ra maintained the discipline of the Arkestra even as he used it to differentiate his organization from other disciplined groups, including the Nation of Islam in the 1950s and the Black Panther Party in the 1970s.[74] In 1971, Ra also began to express this physical sense of discipline in a series of numbered compositions entitled "Discipline [xx]," which he described to Szwed as requiring extremely precise two- to three-note hockets in the Arkestra's horn section; "the slightest variation would destroy the thing," Szwed quotes.[75] Ra's "Discipline" compositions became a mainstay of his live performances with the Arkestra for the rest of the decade.

The physical discipline that Ra demanded of himself and the Arkestra was complemented by a metaphysical discipline. As Jayna Brown has interpreted, Ra's discipline "is essential to be able to face the vastness of the universe. [...] [I]t is required of those who would do nothing less than travel to infinity."[76] In an undated quotation included in Szwed's biography,

Ra elaborated on the ultimate goal of his discipline: *"Discipline ought to permit people to find the most natural things* [...]. Now, in rehearsal we do just like a football team. I give them their little exercises, strict, disciplined rhythms. But when you get on the field you don't do exercises. Then it's part of the game, and you do what you have to."[77]

For Ra, discipline could be located both in training exercises and in the emergent conditions of live performance, when the members of the Arkestra "do what [they] have to." Unlike winning a football game, however, the ultimate goal of the Arkestra's performance was to "find the most natural things," which meant abandoning knowledge and control of one's self and one's instrument in favor of what Daniel Kreiss has described as "the psychological integration of all individuals and the creation of intersubjective awareness."[78] With the Arkestra, Ra sought to remain fully in control of this intersubjectivity. Indeed, immediately after Ra insisted that all of the Arkestra's music, no matter how chaotic, was "written to be" in the program note from his November 1970 performance in Queen Elizabeth Hall, he continued with "Advice to the Arkestra[:] Don't try to count this. If you can't feel it, then it's no good. This is precision music. All parts mesh together. *You play what I write for you.*"[79]

Recordings from this very concert, however, suggest that the Minimoog could not play what Ra wrote for it. Instead, Ra used the instrument to "demonstrate infinity."[80] Ra's simultaneous claims to authority over the Arkestra's performance and his lack of ability to control the Minimoog are resolved by Ra's claim that the Minimoog demonstrates infinity, which cannot be measured or even comprehended. When he played the Minimoog, Ra put himself under a higher, cosmological authority. In the years immediately following his encounter with the Minimoog, Ra articulated this authority in his poetry using many names, such as the "Natural-Master," the "Intergalactic Master," the "Omni-Master," and the "celestial spirit master."[81] Through "Intergalactic master control," Ra wrote in his 1972 poem "Intergalactic Master," one could enter the "full age," a "different connotation / symbolization from the void / immeasurable fullness."[82]

Specifically, Ra manifested this immeasurable symbolization from the void in the form of sounds that escaped extant harmonic conceptions of tone, especially twelve-tone equal temperament. Arkestra bassoonist James Jacson remembers Ra telling him, "You know how many notes there are between C and D? If you deal with those tones you can play *nature*, and nature doesn't know notes."[83] By late 1971, Ra claimed that he had encountered other electronic musical instruments that could play precisely tuned microtonal scales, with eight or sixteen scale degrees between an equally tempered half step.[84] Although the Minimoog could easily produce sounds whose perceived pitch escaped extant understandings of equal temperament, those sounds could not be precisely controlled; its oscillators, which drifted with fluctuations in ambient air temperature and were unable to be synchronized in phase, were controlled by the imprecise sweep of knobs, not precise steps of a microtonal keyboard. This meant that the instrument presented Ra with an apparently infinite spectrum of tones between notes.

Ra's prototype Minimoog was performatively uncontrollable, with a "mind of its own," yet Ra understood that emergent mind, like his own, to exhibit a discipline that transcended earthly ways of being and knowing while remaining opaque. In this sense, Ra's black-boxed, cybernetic musicality—his embrace of unknowable systems that valorized the immediacy of performance over that performance's legibility in extant social, empirical, or aesthetic models of music—differed from the cybernetic musicality of the other musicians in this book. Black boxes such as Buchla's modular electronic music system, Oliveros's oscillator-and-tape delay systems, and Lucier's electroencephalogram-loudspeaker systems suggested visions of hybridized, cybernetic beings. For these musicians, the new subjectivities of these beings produced through cybernetic technoscience retained, by varying degrees, the social identity of "the composer"; their musicality was evidence of their most profound humanity. For Ra, however, the cybernetic model of the black box prompted a different kind of technosocial imagination.

Prior to encountering the Minimoog, Ra had already, for decades, been thinking, writing, and talking about himself, along with other people racialized as Black, as a nonhuman or more-than-human subject. Thus, rather than allow him to unlock the most profoundly human parts of his nature, the Minimoog led Ra to imagine a new kind of subject whose humanity had always-already been called into question. The ontology of this new subject could not be, and indeed refused to be, reduced to extant Western understandings of science or music. Yet at the same time, it engaged in cosmologically attuned "natural" performances. For Ra, the Minimoog suggested an ontology not only of a black box but also of what I provisionally call a Black box, a cybernetic entity whose racialization as Black is brought to bear on its ontology of other-than-humanity, what Ra called the "alter-life."[85]

OPACITY AND THE BLACK BOX

Alterity had long been a concept at the center of Ra's philosophy, which he had been developing since the 1940s. In his early years in Chicago, Ra, along with his spiritual and business partner Alton Abraham, developed an elaborate discourse about the ontology of people racialized as Black, describing them as non-, more-than, or other-than-human with constantly shifting terminology. For example, in a 1951 broadside, Ra and Abraham wrote, "Negroes are not men. Negroes do not belong to the race called man. The truth is that Negroes are human beings, and in trying to be a man they are all out of course."[86] Ra and Abraham attributed this status of being "not men" to a biblical passage: "It is also written concerning Negros: 'they know not, neither do they understand; they walk on in darkness.'" For Ra and Abraham in 1951, Black people—as "human beings" that do not belong to the "race called man"—were defined by their not-knowing, their not-understanding, their existence in darkness.

By the late 1960s, Ra had continued to develop his philosophy of alterity—what he began to call alter-life—by further separating his ontology from mankind on planet Earth, and further identifying with darkness and the unknowable.[87] In a 1968 prose poem, "My Music Is Words," he wrote: "My natural self is not of this world because this world is not of my not and nothingness, alas and happily, at last I can say this world is this unfortunate planet." In a strikingly pessimistic appraisal of earthly life, Ra proposed:

> There are other dimensions and the equation of it is every other world in the infinity of the universe. This is the why of the music I represent; and this is the why of the image of a better world: the alter-life for the alter-life is different from the life of this world. This is a mean world, and by the same token, it is a poor world, poor in spiritual values, void of natural contact with the natural-infinity/otherness Being. [...]
>
> Freedom to me means the freedom to rise above a cruel planet. FREEDOM TO ME MEANS THE FREEDOM TO RISE ABOVE A CRUEL

> PLANET AND TRUE PROTECTION IS PROTECTION BY THE BROTHERS OF NATURE AND NATURES GOD... INFINITY EVERNESS....
> ["]The lifting up shall be the casting down," and it is written that he cast upon them in his anger and fury "a band of evil angels."... My words are the music and my music are the words because it/is of equation is synonym of the Living Being... darkness upon the phase of the deep... the face phase... the eye of infinity... black equation from and of the angelic is... the immeasurable ARE.[88]

For Ra in 1968, the foundation of his subjectivity was infinite, unknowable, and immeasurable, yet also natural, universal, and cosmological. In this same poem, Ra concludes: "The music comes from the void, the nothing, the void, in response to the burning need for something else. And that something else is something else/this nothing, this outer nothing is out of nothing, it is the music of the spheres."[89]

Ra claimed that his music emerged from an absolutely unknowable void that was governed by what Ra called "immeasurable equations," cosmological laws that were demonstrably true and also beyond comprehension. Ra demanded this darkness, this immeasurability, from the members of the Arkestra by asking them to abandon comprehension of what they were playing. (For example, James Jacson recalls Ra saying to him, "Jacson, play all the things you *don't* know! You'll be surprised what you *don't* know. There's an infinity of what you *don't* know.")[90] While Ra created compositions for the Arkestra that would produce such performances of not-knowing, Ra could not play these compositions with the Minimoog. As his performances with the instrument throughout late 1970 and early 1971 suggest, the Minimoog became a model for the cybernetic subjectivity that Ra developed in this period of a racialized Black box, demonstrating an unknowable, immeasurable "black equation from and of the angelic" that insisted on its emergence from "the void, the nothing, the void."

Ra's philosophy of nothingness, infinity, Blackness, immeasurability, and the void resonates with much contemporary Black thought about the question of Black subjectivity in a more general sense. Its questioning of the category of the human recalls Sylvia Wynter's provocation for a "new genre of the human," especially given Ra's similar interest in cybernetic and autopoietic processes.[91] Its pessimism about the nonhumanity of Black people and the impossibility of freedom on Earth, which Ra sometimes articulated with extremely disturbing language, intersects with discourses of Afropessimism, especially Orlando Patterson's concept of social death and Frank B. Wilderson III's position that "human life is dependent on Black death for its existence and its coherence."[92] Indeed, in a 1967 interview with the poet Henry Dumas, Ra explicitly connected Blackness with death: "The blackman [sic] is under the name of death. He carries it around with him. The word *negro*. It is the same as the word *necro*. The g and c are interchangeable according to cosmic mathematics. It is an equation. A negro equals a necro. The sound of one is in the other. In the Greek *necro* means dead body. Necropolis is the city of the dead. Once you accept the name without checking into things you are automatically a citizen of the city."[93]

Yet despite these many intersections, which deserve their own much larger study, here I focus on Ra's claim that his world is one of "not and nothingness," and that his music emerges from "the void." For Ra, his music, like his self, emerges from something radically unknowable and unfathomable. The Martiniquais philosopher Édouard Glissant has also theorized the ontology of the void, and the refusal to be understood, through his concept of opacity as a fundamental characterization of Black life. For Glissant, Black subjectivity emerges from the terrors of the Atlantic passage, "three times linked to the unknown," through the belly of the slave ship, the depths of the ocean, and the unknown destination in front of the ship's bow; though removed by time and geography from these experiences, he argues that the "experience of the abyss lies [both] inside and outside the abyss."[94] Even for the descendants of enslaved people who did not personally experience these abysses, "the unconscious memory of the abyss served as the alluvium" for metamorphoses in their understandings of self.[95] The alluvium at the bottom of the abyss functions to muddy the vision of the Western gaze, disrupting the transparency of the "mirror in which Western humanity reflected the world in its own image. There is opacity now at the bottom of the mirror, a whole alluvium deposited by populations, silt that is fertile but, in actual fact, indistinct and unexplored even today."[96] Opacity, for Glissant, is the ontological condition of the abyssal subject; opacity is "that which cannot be reduced."[97] As John Drabinski argues, the ontological condition of opacity led Glissant to develop what he calls the "aesthetics of an abyssal subject [. . .] tantamount to a defense of the elusive utterance, its opacity and murky depth, and thus a return to the life of contradiction."[98]

The aesthetics of Ra's compositions for the Arkestra certainly reflected the music's emergence from such an abyss; as Fred Moten notes, Glissant specifically invokes the sound of Black utterance as a site of generative opacity. Indeed, describing the strategy of "camouflaging" speech under the conditions of enslavement, Glissant writes: "This is how the dispossessed man organized his speech by weaving it into the apparently meaningless texture of extreme noise."[99] For Moten, however, it is more than Black speech that produces this "real problem and [...] real chance for the philosophy of the human being." It is the "animative materiality—the aesthetic, political, sexual, and racial force—of the ensemble of objects we might call black performances, black history, blackness" that creates openings for the possibilities of human being.[100]

Ra's early exploratory solos with the Minimoog, unlike his compositions for the Arkestra, do not merely illustrate an abyssal aesthetics; they become the loci of what Moten calls "an ongoingly other recording of event, object, music."[101] In these moments, both the Minimoog and Ra himself remained resolutely opaque, avoiding the legibility of musical composition or even improvisation within a composition in favor of a doubly illegible moment of performance—where both Ra and his Minimoog remained black boxes. The consequences of such a refusal for Ra's "philosophy of the human being" were clear. In a prose poem, undated but almost certainly written after Ra's encounter with the Minimoog in 1970, Ra wrote: "I do not remember being born therefore I do not like to speak of birth as a self-authorized known fact. I think I am here but even that thought might be some *pro-creators dream-blueprint designed plan for* cosmo-computer-reference [emphasis in original]. The essence of me is the ETERNAL SPIRIT BEING. My music is like my self: it is not born. IT HAPPENS."[102]

Here, Ra characterizes his subjectivity with cybernetic language, positing a performative ontology that valorizes the performance (the "happening") of his self and music over the knowledge (the "self-authorized known fact") of its birth. The fact that Ra drew so heavily on language from cybernetics and systems theory is not simply a consequence of his working in a moment of heightened public interest in the futuristic "information age";

it is so because he found in that language a set of metaphors that resonated with his concerns about subjectivity, discipline, precision, and freedom.

I bring the metaphor of the black box, and the racialized Black box, to bear on Sun Ra's performances with the Minimoog because they emphasize the ways in which race, sound, and technology contributed to Ra's ontology of alterity. For other musicians studied in this book, black box metaphors were often characterized as liberatory, leading to projections of unbounded personal expression, expansion of the individual mind, and posthuman cyborg musicality. But in the case of Ra and the Minimoog, black and Black box metaphors help us see how Ra reinscribed the dynamic of discipline and precision that he demanded of the Arkestra onto himself, ultimately binding his actions and his very being to an external master. In a 1972 poem entitled "The Universe Sent Me to Converse with You," Ra describes music as "Nature's—Natural-Master's—creation voice," and insists that through his residence in the "Cosmo-usual," rather than on "this planet," "I am doing what I am supposed to do, / I am being what I am supposed to be."[103]

While Ra's reverence toward a cosmological authority could be understood as a deistic expression of faith in the creator of the natural universe, that faith was expressed through a demand for discipline, for putting himself under the control of a cosmological master. The technoscientific language Ra used to describe such control, as Chude-Sokei reminds us, reinscribes a "'grammar' of dominance" that Ra often wielded over the Arkestra, as the architect of a disciplined cybernetic system.[104] In his earliest performances with the Minimoog, Ra was forced to use this same logic on himself. It is not coincidental, then, that the period in which Ra was negotiating his relationship with the Minimoog was also the period in which he began to write and collect the majority of his poems, many of which are cited in this chapter, that articulated his philosophy of alterity, valorizing discipline and denying the possibility of earthly freedom.

CLOSING THE BOX

What did it mean for a musician at the turn of the 1970s to imagine themselves to be a black box? For Ra, the Blackness of this metaphor—its rootedness in racialized constructions of subjectivity, its profound opacity, its challenge to extant forms of knowledge and being—promised a new kind of natural discipline, aligned with the cosmos, that cybernetically swerved from knowledge in favor of performance. Ra's embrace of cybernetic and systems-theoretical language provided him with a matrix of possibility for a new mode of existence—which perhaps paradoxically, or perhaps necessarily, reinscribed the racialized dynamics of control inherent in the cybernetic paradigm. As Kara Keeling has argued, "for Sun Ra, a serious engagement with Black existence leads, on the one hand, to vital articulations of music and sound in the face of various forms of death and, on the other hand, to the quotidian violence that authorizes and reproduces present social relations."[105] Here Keeling specifically refers to Ra's reproduction of settler colonial logics in his fantasy of extraterrestrial space exploration and eventual colonization; she hears in this fantasy a "yearning for another world, another planet that operates according to the space-time of Black liberation." For Keeling, "such a world is not premised on dispossession, ownership, property, and exploitation."[106] Yet despite the absence of those premises, this world would be premised on the concept of discipline: of the Arkestra's submission to Ra's discipline, and of Ra's submission to the cosmological control of an external supernatural being.

Rather than seeking out a liberatory gesture in Ra's embrace of cybernetic ontologies, I argue that it is perhaps more productive to understand Ra's disciplined performances with the Minimoog as what Saidiya Hartman calls "scenes of subjection": "the enactment of subjugation and the constitution of the subject" in spectacularizations of Black life that "illuminate the terror of the mundane and quotidian rather than exploit the shocking spectacle" of violence.[107] For Hartman, scenes such as the singing of a coffle of enslaved people represent such mundane and quotidian terror; they "stress the complexity and opacity of black song."[108] Rather than hear these songs as an "index or mirror of the slave condition," Hartman "emphasizes the significance of opacity as precisely that which enables something in excess of the orchestrated amusements of the enslaved and which similarly troubles distinctions between joy and sorrow and toil and leisure."[109]

Ra's Black-boxed, opaque performances with the Minimoog—which denied the possibility of freedom, insisted on discipline and precision, and spectacularized the condition of being under cybernetic control—similarly troubled distinctions between joy and sorrow, toil and leisure. Inharmonic, noisy, chaotic, and apparently exhibiting extreme discipline, rather than freedom, these performances were quite literally scenes of subjection, of the formation of a new kind of subject. Ra and his Minimoog did not conjure sonic or political freedom; they produced opaque sounds and a politics that saw discipline, precision, culture, and beauty emerging from what Ra called the "kingdom of bondage," referring to ancient Egypt.[110] Deeply pessimistic about Black life on Earth in the twentieth century, Ra instead embraced a discipline and precision he perceived as natural; Ra's concept of nature, in turn, emphasized its emergence from "the void, the nothing, the void" of outer space. With this abyssal origin, Ra's disciplined, cybernetic performances with the Minimoog created an abyssal aesthetics, reveling in their opacity and their irreducibility to language, knowledge, or even music.

While many contemporary interpreters have focused on what Thom Holmes calls the "liberating sound" of Ra's performances with the Minimoog, which in the words of Paul Youngquist produced a "synthesized rainbow of auditory vibrations"—perhaps a subtle nod to the liberal discourse of diversity—I instead argue that the core of Ra's engagement with the Minimoog lay not in its "liberatory" sonic output, but within the disciplined, opaque abyss of the black box itself.[111] Ra's demonstrations of infinity with his Minimoog throughout the early 1970s were not liberatory moments in which Ra as an individual subject, completely in possession of himself, was able to more freely express his internal self through a new electronic musical instrument. Instead, they were moments in which Ra enacted his own subjection to a higher cosmic authority, fantasizing a direct, electronic extension of a subjectivity that transcended the human and indeed human life on planet Earth. In this specific way, Ra's positioning of the Minimoog has something in common with the musician-instrument relationship with which this book began—Morton Subotnick's fantasy of a composer's black box that would allow him to immediately extend his inner subjectivity through electronic sound. But what Ra extended through his performances was not a transparent expression of his own subjectivity; it was an opaque impression of an alter-life that was not entirely his own. In their opacity and their alterity, Ra's performances with his prototype Minimoog articulated, in his own words, "a different order of sounds synchronized to the different order of Being."[112]

Afterword

WHAT CAN A COMPOSER DO?

This book began by asking the question: What can a black box do? Electronic musical instruments provoke this question quite easily. Their interfaces, whether simple or complex, seduce the user into an exploratory, experimental mode of playful interaction. In contrast with an acoustic musical instrument, a black-boxed electronic instrument's behavior depends on its internal configuration, which produces a cascading series of related questions: What do these knobs do? What do those lights mean? What happens if I connect these two points together—what do the lights mean and the knobs do then? The answers to these questions, to musicians, often emerge as sounds; they resist static explanation, control, or knowledge, and instead swerve toward immediacy, chaos, and ineffability. (What musician has ever read the user manual first?) Sounds seem to connect these black-boxed, electronic instruments to humans by somehow communicating with an apparently deep auditory process embedded in human consciousness that can be described with the same scientific terms of frequency and amplitude. Playing with musical black boxes can easily produce a sense of wonder, discovery, and enjoyment that specifically results from the box's opacity, its refusal to give up its secrets even as it reveals itself through sound.

It is easy to imagine "sound" as a neutral vibratory current, a signal that flows in and out of such boxes, swirling between human, instrument, air, tape, and even digital audio code.[1] But I think that attending solely to the *sound* of a human/black-box encounter masks another important question that needs to be asked the other way around: In an encounter between a human and a musical black box, what can a human musician do? That is to ask: How does a human generate music with such an instrument? How does it come up with ideas for how to use musical instruments that refuse to be understood? What can a human musician do with a black box, together, in the here and now? To ask this question the other way around is not to valorize nor devalorize the human, nor to shift the locus of agency onto or away from technological or material actors; it is not to insist that instruments have agency, or even that sound itself has agency. Rather than opening a window onto an ostensibly nonhuman domain of sound for humans to perceive, I hope to have shown that the black boxes followed throughout this book instead held up a mirror to their human users, forcing them to reevaluate what constitutes a musicking human in the first place. If the composer's black box is controlled by the composer as black box—if it's black boxes all the way down, in an increasingly informational and cybernetic world—then what is a composer, anyway?

In this book's introduction I discussed a common response to this question: that in the information age a composer, just like many other human subjects, is always-already posthuman. The moment that a composer such as Morton Subotnick imagined that an electronic, computational black box could extend his innermost self through pure electricity; the moment that Buchla imagined his Box as a psychedelic, computational model of his own mind; that Pauline Oliveros imagined that her body, patched into her "very complex nonlinear music making system," could "do the work" of improvisation automatically; the moment that Alvin Lucier entertained the notion that the electrical activity of his brain could "play" percussion instruments; the moment Sun Ra speculated that his entire being could be "some pro-creators dream-blueprint designed plan for cosmo-computer-reference"—these are all moments at which humans have been, as Katherine Hayles wrote in 1999, "put [...] into a cybernetic circuit that splices [their] will, desire, and perception into a distributed cognitive system in which represented bodies are joined with enacted bodies through mutating and flexible machine interfaces."[2]

But as I have shown in this book, none of these moments happened without earlier, messier, more ambiguous encounters with musical black boxes. It took Subotnick six years of working with theatrical enclosures and reading Marshall McLuhan before he came to the idea of a composer's black box; the countercultural fantasy of Buchla's Box was premised on the technocultural Cold War fantasies of the military-industrial-academic complex in which he was embedded; Oliveros came to her oscillator-and-tape-delay system through years of thinking about improvisation and her own body; Lucier's other early works with black boxes valorized the ambiguities they produced; Ra could not master his Minimoog like he mastered his other instruments. By the mid-1970s, perhaps, the condition of posthumanity, variously disembodied (Hayles) or hyperembodied (Eshun), was common among all of the musicians studied in this book, and many more.[3] But in the 1960s, as these musicians were first encountering electronic audio technologies and assembling them into black boxes for experimentation, that condition was not yet posthuman, but rather something else—something that allowed for more differentiation.

That something else was not often legible as music in the 1960s. Indeed, as Pauline Oliveros's friend Stuart Dempster characterized it, early encounters with such black boxes often produced what he called "garbage."[4] But as I suggested in chapter 3, it was precisely this "garbage" that bore the traces of some of the most interesting musical potentials to emerge from human/black-box interactions in the 1960s. The many reels of tape that Oliveros recorded of her earliest real-time improvisations with oscillators; the sketches that Lucier saved of his contradictory, pluralistic, hungry negotiations with cybernetic technoscience; the bootlegged live recordings of Sun Ra's meandering Minimoog solos from his 1970 tour: these are the artifacts of black-boxed musicality before it had time to settle into more familiar patterns and schemas of posthuman musicality. I chose to focus on these artifacts because they represent a brief but consequential moment in these musicians' careers, and perhaps also in the larger history of American technoscience, when the incorporation of information theory and cybernetics into everyday life was still being negotiated and had not yet become naturalized.

After this period cybernetics was both everywhere and nowhere. In 1988, Subotnick could claim that he had fantasized a music easel that would function as a composer's black box thirty years earlier without comment from an otherwise extremely attentive interlocutor.[5] By the early 1970s, Oliveros had abandoned recording her interactions with electronic black boxes in favor of developing protocols for the exploration of the human black box, which she playfully called "software for people" in 1980.[6] By that same year, Lucier had withdrawn his ambiguous early works such as *Music for Amplified Lip*, and performances of *Vespers* began to feel like earnest scientific experiments. And Sun Ra quickly abandoned the prototype Minimoog Model B in favor of a more stable Model D (as seen in his 1974 film *Space Is the Place*) before entirely switching over to polyphonic synthesizers that behaved more like electric organs than black boxes, at every opportunity emphasizing that his performances were disciplined and precise.[7]

Before all of this shoring up, this adjustment to a new national (and global) technoscientific environment, it was not so clear what the increasing ubiquity of cybernetics and information theory would, or could, mean for music—a cultural endeavor that has often been mobilized as the most profound evidence of humanity, something a machine could never fully embody. But during the period of the 1960s studied in this book, when black-box thinking and black-box technology began to draw closer and closer connections between human and machinic bodies, a wide spectrum of possible configurations and consequences emerged before it was quickly subsumed by the mundane normalcy of the information age.

EX-COMPOSITION

Here I return to the concept of *ex-composition*, which I began developing in chapter 2, to name one of the major ways that this black-boxed musicality in the 1960s differed from later relational dynamics between humans and informational instruments. With this term, I wish to clarify a subtle conceptual shift in the understanding of the human production of music from a nineteenth-century notion of authorship to a more distributed, cybernetic coproduction of musical phenomena, before that coproduction became reinscribed as the composition of musical "works":

scores and recorded performances intentionally, proudly authored by the humans involved in such human/black-box pairings (even if, and especially when, some amount of authorial agency was assigned to black boxes as "composers" themselves). Importantly, I am not claiming that such ex-compositional musicality is ahistorical, universal, or natural; what I am claiming, on the contrary, is that it is specific to the experiences of humans who for the first time encountered black-boxed electronics and began to dance, negotiate, explore, work, think, and dream with and through them, especially in America in the 1960s, as these technologies first began to filter out of the military-industrial-academic complex and into public life.

As discussed in chapter 2, I came to ex-composition from what the psychedelics researcher Richard Alpert called "ecstatics," his provisional (and short-lived) scientific approach to the "systematic measurement, description, and production of the ecstatic state—that is, the expansion of consciousness."[8] For Alpert, the ecstatic state was desirable, and productive, precisely because it placed human consciousness "beyond learned modes of experience (particularly the learned modes of space-time-verbalization-identity)." In 1964, Alpert articulated ecstatics in explicitly information-theoretical and cybernetic terminology—that is, that an ecstatic state is one in which the human brain is forced, through various techniques, to increase the absolute number of possible electrical connections between neurological synapses, thus eliminating the "limits of learned cultural programs." In this definition, ecstatics expands human consciousness, producing a more capacious sense of individual self in the ecstatic subject.

In 1970, after many such ecstatic experiences (achieved through LSD, hypnosis, meditative states, and other technologies), and after his spiritual rebirth and transformation into Ram Dass, Alpert/Dass began to more explicitly associate the ecstatic state with the dissolution of human subjectivity. This would first begin with the quieting of its Freudian ego, as he wrote in his widely celebrated book *Be Here Now*:

> When you are with a candle flame you ARE the candle flame
> and when you are with another being's mind you ARE the other being's mind
> when
> there is a task to do
> you ARE the task
> the mindless quality of total involvement that comes only when the Ego is quiet and there is no attachment.[9]

Although this passage suggests the expansion of an extant human subject (albeit one without ego), strongly recalling Norbert Wiener's 1950 poetic aside that "the individuality of the body is that of a flame rather than that of a stone," Dass quickly went one step further: the ego needed not only to be quieted, but also killed. "Do you realize when you go on that journey in order to get to the destination YOU can never get to the destination? In the process YOU must die. MUST DIE. Pretty fierce journey pretty fierce requirement / WE WANT VOLUNTEERS," he wrote in 1970.[10] Although Alpert's stance was not yet as extreme in early 1966, when he commissioned Buchla to create a system for his friends the Anonymous Artists of America, the kernel was there; a psychedelic experience was valuable insofar as it enabled new ways to experience the world, and new ways of understanding one's identity in relation to that expanded world, through inexorably changing (and potentially destroying) the subjectivity of the human being.

The transformation of the human subject was not a mere philosophical exercise. Along with the AAA, Buchla, and many others in the mid-1960s Bay Area, Alpert had easy access to powerful chemical and electronic technologies that produced real, immediate effects on human sensation and cognition. This ecstatic experience was, I argue, one of the principal effects of working with a musical black box such as the Buchla Box, which is perhaps why Alpert donated one along with a massive amount of LSD to the AAA as an inaugural gift. Getting a black box such as the Buchla Box to make music in a traditional sense—whether rock and roll, in the case of Len Frazer in the AAA, or "chamber music 20th-century style," in the case of Subotnick—was nearly impossible. Recall the "desultory swooshing sounds" that sounded great to Frazer but annoyed the AAA's bandleader, or the thirteen months it took Subotnick to figure out how to produce *Silver Apples of the Moon* with his Buchla system. Thirteen months is a far cry from the immediacy of the ecstatic, psychedelic effect that such an instrument had on its user. As Buchla wrote in a 1970 poem, the only recorded articulation of his philosophy of technology, his instruments allowed humans to "listen to the results of the / idea in real time corrective / feedback in one biochemical / response cycle on line / realization of ideation."[11]

Buchla's embrace of an "on line realization of ideation" through the use of his Box, a cybernetic, informational electronic musical technology, was shared by musicians who never went near illicit drugs. The "on-line computational thinking" that Alvin Lucier sought to execute in *Vespers*, for example, expanded its human performer's cognitive and sensory capacities by forcing them into an ecstatic state through the use of Sondols and a strict behavioral protocol. The ecstatic state achieved in *Vespers*, like the ones described by Alpert and Buchla, emerged from ostensibly forcing the human mind to increase its computational capacity; like Buchla, Lucier conceptualized the human being as a human biocomputer. This was not coincidental. Both Buchla and Lucier borrowed this metaphor from John C. Lilly, one of the most (infamously) popular public cybernetic thinkers in the 1960s. Both picked up on a psychedelic streak in Lillean thought which, while still relatively nascent in the 1960s, would become a full-blown public embrace (and perhaps a public embarrassment) of acid-fueled theorizations about human and nonhuman (both animal and extraterrestrial) consciousnesses well into the 1970s and beyond.[12]

The computational kernel of the ecstatic metaphor, of the human as black-boxed biocomputer that could ecstatically stand outside of itself and change its own programming, also made it to Oliveros's thought by the 1970s. In a 1973 article for the short-lived publication *NuMus West*, she mused: "How have I programmed my Bio-Computer?"[13] The reprogramming of Oliveros's biocomputer through early ecstatic experiences with black boxes changed her conception of self; ex-composition seems an apt metaphor for her work both during this period and immediately thereafter, since beginning in 1971 Oliveros explicitly claimed to have "abandoned composition/performance practice as it is usually established today."[14] This proclamation came in the first publication of her *Sonic Meditations*, a collection of practices Oliveros had developed with the ♀ Ensemble in San Diego, a gynocentric consciousness-raising group of women, mostly without musical training, who met weekly to conduct what Oliveros called "Sonic Explorations." As she would write in the introduction to a 1974 collection of the *Sonic Meditations*, these explorations posited the existence of "Sonic Energy," which allowed for "communication among all forms of life" and which could be "trasmit[ted] within groups" to facilitate healing. Implying that sound behaved like an informational signal that could be communicated or transmitted, Oliveros's *Sonic Meditations* were protocols for humans to stand ecstatically outside themselves and look inward into their own black-boxed subjectivities, encouraging participants to explore "auditory fantasy" and "auditory memory" through the use of "trigger questions," technologies for allowing the subject to hear within herself in novel ways.[15]

The ecstatic state was also at the heart of Sun Ra's performance practice with his prototype Minimoog. Notably, Ra did not, and perhaps could not, use this instrument either to compose music for his Arkestra or to perform his extant compositions; its opacity led to an unprecedented and unique musical practice within Ra's long career during his tours of Europe and Africa in 1970 and 1971. I find ex-compositional to be a fitting description for Ra's performance practice with the Minimoog on these tours because it performed the refusal of composition, of the legible organization of sound, in real time. Out of all of the musicians brought together by this book, Ra perhaps articulates the ex-compositional most immediately. In the ex-compositional, musical agency does not emerge entirely from an instrument, nor does it emerge from a human being that is extended, prostheticized, or otherwise enhanced by an instrument. It essentially abandons the idea of compositional agency in favor of the immediacy of emergence, of emergency. The ex-compositional has something to do with improvisation, but an improvisation that swerves from mastery because it remains opaque. It does not entirely give up the human, yet it does not reinforce extant conceptions of humanity. Perhaps, if anything, it abandons the idea that music is somehow essential to human subjectivity, agency, and freedom as we know it. This may at first seem like a closing off of one of the most widely celebrated activities of human cultural expression; but listening back to the ex-compositional music making of people like Subotnick (despite his claims to the contrary), Buchla, Oliveros, Lucier, and Ra doesn't feel like a closing. It feels like an opening.

THE COMPOSER'S BLACK BOX, SIXTY YEARS LATER

What has become of that opening? In 2025, as in the 1960s, black boxes appear to be everywhere; in many respects, they have become the standard systems through which most music is made, recorded, and distributed. Not even accounting for the fact that digital audio workstations are the default technology for producing music, or the fact that algorithmic recommendation software and streaming platforms are the default technologies for distributing music, the market for electronic musical instruments has also defaulted to the black box.[16] For over thirty years we have been in the midst of what Trevor Pinch and Frank Trocco have dubbed the "analog revival," a surge of interest in early electronic synthesizers that has produced (or perhaps renewed) robust techno-utopian fantasies about these instruments' promises for individual self-expression.[17] That revival has created a commercial synthesizer industry that imagines Buchla and Moog as its two patriarchal "fathers," with year-on-year growth that has led to increasing industry consolidation and capital investment.[18]

The current state of musical black boxes is a product of both the utopian promises imagined by musicians through their initial encounters with such black boxes in the 1960s and the failures baked into those promises, their utopian impossibility. Even as these black boxes amplified their human users' social and technological fantasies of immediate musical creativity, unfettered access to the world of sound, or even a musicality beyond the human, they always recursively sabotaged such promises, setting up an impossible stage on which those fantasies could be played out again and again through musical practice. These impossible promises are in a sense a microcosm of what the historian Frederic Jameson somewhat glibly termed his "'unified field theory' of the 60s, [...] a single process at work in first and third worlds, in global economy and in consciousness and culture, a properly *dialectical* process, in which 'liberation' and domination are inextricably combined."[19] Rather than simply celebrate the techno-utopian fantasies of the musicians studied in this book or simply critique them for the near-revanchist attitude toward music that they later enabled, I posit the *ex-compositional* to point to both the ways that black boxes potentially dissolved the subject of the composer and the ways that they kept that social identity around, as a specter that haunted the moments of ecstatic uncertainty produced through black-box encounters.

In the current moment, the valence of those ex-compositional moments of ecstatic uncertainty—the material consequences they could have for a musician—bear that same dialectical combination of liberation and domination. As Jameson presciently predicted, since the 1980s, we have been in "an effort, on a world scale, to proletarianize all those unbound social forces which gave the 60s their energy."[20] As musical black boxes have accumulated more capital (in the form of industrial production and institutional funding), they have transformed the potentially consciousness-expanding, ego-killing, ex-compositional moments of black-box musicality into moments that fuel the booming synthesizer industry, of the endless designing, manufacturing, selling, buying, and trading of black-boxed instruments that continue to make impossible techno-utopian promises. While these moments still have the potential to make their users question their ontological or epistemological assumptions about sound, humanity, the human body, human cognition, or music, they also sell products, and produce sounds that are now thoroughly legible within academic and commercial institutions. A Buchla Box is no longer a rare psychedelic technology that makes no sense in existing social worlds of music; it is now a "tool" to be used in the service of music production within institutional and commercial landscapes that remain fundamentally unscathed by the challenge of the black box.[21]

Where do we go from here? There is perhaps still something recoverable in the moment of black-box encounter, especially in the moment of first encounter, when the box still retains some degree of opacity. The scene of a human's first encounter with a musical black box reminds me of Theodor Adorno's famous image of a child sitting at the piano, which, as Seth Brodsky reminds us, went through several revisions between the 1930s and 1960s while Adorno refined his position; at first purely negative, Adorno's musico-political imaginary later became utopian. "The child searching for a melody on the piano offers the paradigm of all true composing," Adorno wrote in 1929. "The composer searches for what perhaps was always already there [...] on the undiscriminating black and white keys of the keyboard from which he must make his choice."[22] For Adorno in 1929, the composition of actually new music was impossible, precisely because the composer's instrument remained the same as it had been for centuries. By 1966, however—contemporaneous with the period studied in this book—this impossibility became a utopian longing:

> The new is the longing [*Sehnsucht*] for the new, hardly the new itself, from which everything new suffers. What feels itself to be utopia remains a negative of what exists, to which it remains in bondage. At the center of contemporary antinomies is this: art must be utopia, and wants to be ... at the same time art, in order not to betray utopia to appearance and consolation,

does not allow itself to be utopia. If the utopia of art fulfilled itself, it would be the temporal end of art.... Only through its absolute negativity does art pronounce the unpronounceable: utopia.[23]

In some senses, the musical black boxes brought together in this book did allow for glimpses of utopia, if only momentarily. In front of a piano keyboard, Adorno wrote in 1966, "the chord was always already there, the possibility of combinations is limited, actually everything already rests within keyboard itself."[24] Musical black boxes swerved from the domination of music's cultural technologies—not only the material apparatus of the keyboard but also the theoretical apparatus of pitch, the (psycho)acoustic apparatus of harmonicity, and the social apparatus of the bounded, individuated composer who had mastered the composition of tonally moving forms.[25] With a musical black box there were no chords to be found, the possibility of combinations approached the infinite, and nothing seemed to rest within the instrument itself, since the instrument presented itself as fundamentally opaque.

What did it mean that the liberatory promise of such black boxes—the pathway they presented out of domination—was opaque? Opacity refuses to be legible; it refuses to be understood. It has the potential to refuse, as well, the incorporation of such ex-compositional musical activities into a late capitalist landscape in which musical black boxes become salable commodities, tools purchased by individual consumers for the apparent expression or sonic emancipation of the individuated self. Of all the musicians brought together by this book, Sun Ra perhaps most honestly addressed the contradictory promise of liberation presented by such black boxes. Ra's pessimism about freedom on Earth resonates with Adorno's negative definition of utopia; as Fumi Okiji writes of the practices of other jazz musicians in the 1960s, "the flashes of eschatological utopia that we hear in the sociomusical play take us, momentarily, into a blackened atmosphere 'beyond space, time, causality, individuation.'"[26] While Okiji locates such a blackened atmosphere within jazz in a general sense, I argue that Ra's doubly opaque sociomusical play with his musical Black box, perhaps even along with the sociomusical play of the other musicians brought together in this book, "gives us access to a conflicted subject that will not cohere but rather is in a state of constant rejuvenation through the unstable, generative relations of its disparate ways," as Okiji writes.[27]

Rejuvenation might ultimately be the best potentiality that musical black boxes present their human users. Not as a simple return to childlike innocence, or as a turn away from the social and political violences of the world, but rather as a return to a human subjectivity still in the state of formation within that world, negotiating the contours of its sensation, embodiment, and agency. To ask what a human musician can do faced with a musical black box is to ask after the human in a more general sense; considering the opacity both of the box and of the human gives us a "chance," as Fred Moten puts it, for the "philosophy of the human being."[28] All of the musicians brought together in this book underwent transformations through their black-box encounters; the other side of those encounters led them down differential, diffracted pathways of exploring the possibilities of human musicality, and human subjectivity, that we can perhaps still hear, and experience, today.

Notes

INTRODUCTION

1. "Young People's Concert: 'Bach Transmogrified,'" CBS, broadcast April 27, 1969, posted April 26, 2018, https://youtu.be/J8sS5NkADBE?si=84SXPg2eA_4RbVBD.
2. Moog, quoted in Tam Fiofori, "Moog Modulations: A Symposium," *DownBeat*, July 23, 1970, 14.
3. Moog, quoted in Fiofori, "Moog Modulations," 14. Curiously, despite using this exact same wording for its title, Paul Théberge's *Any Sound You Can Imagine: Making Music/Consuming Technology* (Wesleyan University Press, 1997) does not cite this interview, instead citing the ubiquity of the phrase in the marketing materials of keyboard synthesizers in the 1970s. See Théberge, *Any Sound*, 76.
4. Moog, quoted in Fiofori, "Moog Modulations," 16. Note that use of brackets around ellipses indicates that the author has deleted some words from the quotation. Ellipses without brackets are present in the quoted material itself.
5. Swansen, quoted in Fiofori, "Moog Modulations," 16.
6. *Subotnick: Portrait of an Electronic Music Pioneer* (Waveshaper Media, 2017).
7. Ra, quoted in Fiofori, "Moog Modulations," 34, 39.
8. Ra, quoted in Fiofori, "Moog Modulations," 34, 39.

9. Curtis Roads and Morton Subotnick, "Interview with Morton Subotnick," *Computer Music Journal* 12, no. 1 (1988): 9–18, 12. Although this term was printed as *blackbox* in this interview, I retain the more common *black box* orthography throughout this book. Subotnick also uses the term *black box* in an interview with David Bernstein and Maggi Payne published in David W. Bernstein, *The San Francisco Tape Music Center: 1960s Counterculture and the Avant-Garde* (University of California Press, 2008), 114.

10. Subotnick, quoted in Bernstein, *San Francisco Tape Music Center*, 114.
11. Subotnick, quoted in Bernstein, *San Francisco Tape Music Center*, 114.
12. Subotnick, quoted in Bernstein, *San Francisco Tape Music Center*, 114.
13. Ronald R. Kline, *The Cybernetics Moment, or Why We Call Our Age the Information Age* (John Hopkins University Press, 2017), 5.
14. Kline, *Cybernetics Moment*, 5. See also David E. Nye, *America as Second Creation: Technology and Narratives of New Beginnings* (MIT, 2004).
15. Ra quoted in Tam Fiofori, "Sun Ra's African Roots," *Melody Maker*, February 12, 1972, 32.
16. Creative Audio Archive, Experimental Sound Studio, Chicago, IL, SR-R073, Reel 46.
17. Philipp Von Hilgers, "The History of the Black Box: The Clash of a Thing and Its Concept," *Cultural Politics* 7, no. 1 (March 1, 2011): 46.
18. Norbert Wiener, *Cybernetics: Or Control and Communication in the Animal and the Machine*, 2nd ed. (MIT Press, 1961), 11.
19. Peter Galison, "The Ontology of the Enemy: Norbert Wiener and the Cybernetic Vision," *Critical Inquiry* 21, no. 1 (October 1994): 229.
20. Galison, "Ontology of the Enemy," 229.
21. Kline, *Cybernetics Moment*, 7.
22. Wiener, *Cybernetics*, 14.
23. Geof Bowker, "How to Be Universal: Some Cybernetic Strategies, 1943–70," *Social Studies of Science* 23, no. 1 (1993): 116,
24. Christina Dunbar-Hester, "Listening to Cybernetics: Music, Machines, and Nervous Systems, 1950–1980," *Science, Technology & Human Values* 35 (2010): 132.
25. Comments by Morton Subotnick at *Screening of Subotnick: Portrait of an Electronic Music Pioneer* (Waveshaper Media, 2022), Brooklyn College, Brooklyn, NY, March 22, 2024.
26. Part of Meyer-Eppler's experience in studying phonetics and electrical engineering came from his experience working for the Nazi Kriegsmarine throughout 1943–45; he was categorized as a Category IV Mitläufer after the war. See Jennifer Iverson, *Electronic Inspirations: Technologies of the Cold War Musical Avant-Garde*, New Cultural History of Music (Oxford University Press, 2018), 29.
27. Eamonn Bell, "Cybernetics, Listening, and Sound-Studio Phenomenotechnique in Abraham Moles's *Théorie de l'information et Perception Esthétique* (1958)," *Resonance* 2, no. 4 (December 1, 2021): 523–58.

28. Bell, "Cybernetics, Listening," 11; Lejaren Hiller and Leonard M. Isaacson, *Experimental Music: Composition with an Electronic Computer* (McGraw-Hill, 1959).

29. Brian A. Miller, "Leonard Meyer's Theory of Musical Style, from Pragmatism to Information Theory." *Resonance* 2, no. 4 (December 1, 2021): 475–502.

30. Eric Drott, "Music and the Cybernetic Mundane," *Resonance* 2, no. 4 (December 1, 2021): 586.

31. Harry Olson and Herbert Belar, quoted in Drott, "Music and the Cybernetic Mundane," 587.

32. See Dunbar-Hester, "Listening to Cybernetics," for a survey of multiple typologies of translations between cybernetics and music.

33. Christopher Haworth, "Music and Cybernetics in Historical Perspective." *Resonance* 2, no. 4 (December 1, 2021): 463.

34. Haworth, "Music and Cybernetics in Historical Perspective," 463.

35. Drott, "Music and the Cybernetic Mundane," 581.

36. V. Vale and Andrea Juno, *Incredibly Strange Music*, vol. 2. (RE/Search Publications; Subco, 1993), 200.

37. See You Nakai, *Reminded by the Instruments: David Tudor's Music* (Oxford University Press, 2021).

38. Gordon Mumma and Michelle Fillion, *Cybersonic Arts: Adventures in American New Music*, Music in American Life (University of Illinois Press, 2015). See also Robert Ashley, "The Wolfman, for Amplified Voice and Tape," in *Source: Music of the Avant-Garde, 1966–1973*, ed. Larry Austin, Douglas Kahn, and Nilendra Gurusinghe (University of California Press, 2011), 143–45; and David Behrman, "Wave Train," in *Source*, ed. 108–15.

39. See chapter 4 of this book.

40. David Rosenboom, "Saturation in Multi-Media," *Continuum*, September 1968.

41. After the 1970s, cybernetic metaphors became nearly ubiquitous in music around the globe, even as their meanings moved further away from their specific and narrowly defined scientific use. Christina Dunbar-Hester has commented on the difficulty in establishing concrete connections between musicians and cybernetics: despite the "paucity of primary sources that explicitly linked cybernetics to experimental and electronic music," she noted that in her research, conducted in the late 2000s, she often encountered musicians' "rhetoric [that] seemed quite in keeping with cybernetic theories, but I could not find an explicit link between the source material and people or ideas that were known to be 'cybernetically active.'" Dunbar-Hester, "Listening to Cybernetics," 130.

42. See, for example, George E. Lewis, "Too Many Notes: Computers, Complexity and Culture in Voyager," *Leonardo Music Journal* 10, no. 1 (December 1, 2000): 33–39; George E. Lewis, "Interacting with Latter-Day Musical Automata," *Contemporary Music Review* 18, no. 3 (January 1999): 99–112.

43. Bowker, "How to Be Universal," 110, quoting Arturo Rosenblueth, Norbert Wiener, and Julian Bigelow, "Behavior, Purpose and Teleology," *Philosophy of Science* 10, no. 1 (1943): 18–24.

44. Bowker, "How to Be Universal," 114.

45. Quoted in Ben Kettlewell, *Electronic Music Pioneers* (ProMusic Press, 2002), 143–44. Ciani's comparison of the Buchla Box as cybernetic system to a human being was influenced by her involvement with Erhard Seminars Training (EST), founded by the charismatic leader Werner Erhard in 1971, which many place within the human potential movement—itself deeply invested in cybernetic metaphors. In a 2016 interview with Morton Subotnick, Ciani explained what she learned from the seminar, which was that "human beings are machines. They have their operating systems. If you were given the tools to impact that human technology, you would make progress. In the end I said, 'Oh my God, I'm in love with the machine. Humans are just machines, so I'm okay.'" Morton Subotnick and Suzanne Ciani, "Music Pioneers Suzanne Ciani and Morton Subotnick in Conversation," interview by Frosty, June 24, 2016, Red Bull Music Academy, http://daily.redbullmusicacademy.com/2016/06/encounters-suzanne-ciani-morton-subotnick.

46. N. Katherine Hayles, *How We Became Posthuman Virtual Bodies in Cybernetics, Literature, and Informatics* (University of Chicago Press, 1999), 2–3.

47. Norbert Wiener, *The Human Use of Human Beings: Cybernetics and Society* (1950–1954; Discus Books, 1967), 137–39.

48. Hayles, *How We Became Posthuman*, 1.

49. Alexander G Weheliye, "'Feenin': Posthuman Voices in Contemporary Black Popular Music." *Social Text* 20, no. 2 (2002): 23. Weheliye later expanded this article into a book, *Feenin: R&B Music and the Materiality of BlackFem Voices and Technology* (Duke University Press, 2023).

50. Weheliye, "'Feenin'," 23.

51. Weheliye, "Feenin'," 39.

52. Kodwo Eshun, *More Brilliant Than the Sun: Adventures in Sonic Fiction* (Quartet Books, 1998), 112.

53. Eshun, *More Brilliant Than the Sun*, 79, 80.

54. G. Douglas Barrett, *Experimenting the Human: Art, Music, and the Contemporary Posthuman* (University of Chicago Press, 2023), 2.

55. Barrett, *Experimenting the Human*, 14, 7.

56. Barrett, *Experimenting the Human*, 14.

57. Hayles, *How We Became Posthuman*, 3; Eshun, *More Brilliant Than the Sun*, -002.

58. Allen Strange, *Programming and Meta-Programming in the Electro-Organism: An Operating Directive for the Music Easel* (Buchla & Associates, 1974).

59. Andrew Pickering, *The Mangle of Practice: Time, Agency, and Science* (University of Chicago Press, 1995), 22.

60. Comments made by Subotnick at a public event at the Library of Congress, December 5, 2024.

61. Subotnick continued to work with Buchla to develop friendlier human-instrument interfaces, such as an envelope tracker that could translate the amplitude of Subotnick's voice into a control signal, and other modules in Buchla's 200 series systems. Ra stopped using the Minimoog in favor of other polyphonic electric keyboard instruments, such as the Farfisa organ and the Hohner clavinet; while Ra did play electronic synthesizers well into the 1990s (such as the Crumar Mainman and even a Yamaha DX-7), these were all polyphonic instruments that behaved like electric organs.

62. Coxe began to develop "the Simeon" and changed his band's name to Silver Apples in 1967, the same year as Subotnick's *Silver Apples of the Moon* was released; the two both lived and worked in downtown New York City. However, Coxe claims that he had no knowledge of Subotnick, and that he came to the name Silver Apples entirely independently, from a teenage love of Yeats's poem "The Song of Wandering Aengus." Shane Christmas, "Silver Apples Interview," Perfect Sound Forever, October 2000, https://www.furious.com/perfect/silverapples.html.

63. On the liner notes to *Rip, Rig, and Panic*, Kirk writes: "When they hear me doing things they didn't think I could do they panic in their minds. They all say, 'Well, I didn't know this kind of thing could happen.' Actually, I was doing some things like this when I was in Ohio, but I lost work, because people didn't want to hear this kind of thing." Roland Kirk, *Rip, Rig, and Panic* (Limelight, 1965).

64. It's not clear where this widely circulated quote originated; some have attributed it to Robin Williams, who lived out that decade as a teenager and incorporated this line into his stand-up comedy routine in the 1980s, but others attribute it to either Paul Kanter or Grace Slick, two members of Jefferson Airplane.

CHAPTER 1. MORTON SUBOTNICK
AND THE COMPOSER'S BLACK BOX

1. Morton Subotnick, *Silver Apples of the Moon* (Nonesuch H-71174, 1967).
2. Subotnick, *Silver Apples of the Moon*.
3. Subotnick, *Silver Apples of the Moon*.
4. *Subotnick: Portrait of an Electronic Music Pioneer* (Waveshaper Media, 2017).
5. Curtis Roads and Morton Subotnick, "Interview with Morton Subotnick," *Computer Music Journal* 12, no. 1 (1988): 12; comments by Morton Subotnick at Screening of *Subotnick: Portrait of an Electronic Music Pioneer* (Waveshaper Media, 2022), Brooklyn College, Brooklyn, NY, March 22, 2024.

NOTES

6. Myron Barnett, "Interview with Ramon Sender, Tony Martin, Morton Subotnick," University of Cincinnati College-Conservatory of Music, WGUC-FM, Cincinnati, July 13, 1964, Center for Contemporary Music Archive at Mills College, CCM CD1.

7. Sender quoted in Robert Commanday, "Composer's Forecast: A New Kind of Artistic Man," *San Francisco Examiner*, April 12, 1965, POP-UCSD, box 25, folder 14; see also Ramon Sender, "The San Francisco Tape Music Center: A Report," in *The San Francisco Tape Music Center: 1960s Counterculture and the Avant-Garde*, ed. David Bernstein (University of California Press, 2008), 42–46.

8. Myron Barnett, "Interview with Ramon Sender, Tony Martin, Morton Subotnick," University of Cincinnati College-Conservatory of Music, WGUC-FM, Cincinnati, July 13, 1964, Center for Contemporary Music Archive at Mills College, CCM CD1.

9. HKW Years of Now, "The Technological Big Bang: Tape Recorders, the Transistor & the Credit Card/Morton Subotnick," December 7, 2017. https://www.youtube.com/watch?v=Nvw4Z3QozaY&ab_channel=HKW100YearsofNow.

10. As Subotnick remembered it, Milhaud commented, "Thank you my dear, it reminds me of the old days!" Morton Subotnick, "Major Figures in American Music," interview by Greg Bendian, March 23, 1983, Yale Oral History of American Music, Yale University Libraries, 38; Morton Subotnick, "Morton Subotnick," interview by Todd L. Burns, Red bull Music Academy, 2011, https://www.redbullmusicacademy.com/lectures/morton-subotnick-2011.

11. Nathan Rubin, *John Cage and the Twenty-Six Pianos of Mills College: Forces in American Music from 1940 to 1990, a History* (Sarah's Books, 1994), 86.

12. Martin Brody, "The Enabling Instrument: Milton Babbitt and the RCA Synthesizer," *Contemporary Music Review* 39, no. 6 (November 1, 2020): 783–84.

13. "Princeton Seminars (1959 &1960)," accessed August 28, 2024, https://frommfoundation.fas.harvard.edu/princeton-seminars.

14. Brody, "Enabling Instrument," 778.

15. Milton Babbitt, "Past and Present Concepts of the Nature and Limits of Music (1961)," in *The Collected Essays of Milton Babbitt*, ed. Stephen Peles et al. (Princeton University Press, 2003), 83–84.

16. Babbitt, "Past and Present Concepts," 84.

17. Brody, "Enabling Instrument," 787.

18. Carlton Gamer, "Milton at the Princeton Seminars: A Remembrance," *Perspectives of New Music* 49, no. 2S (2012): 361.

19. Gamer, "Milton at the Princeton Seminars," 361.

20. Gamer, "Milton at the Princeton Seminars," 362.

21. Gamer, "Milton at the Princeton Seminars," 362.

22. Milton Babbitt, "An Introduction to the R. C. A. Synthesizer," *Journal of Music Theory* 8, no. 2 (1964): 263; Babbitt, "Past and Present Concepts," 84.

23. Brody, "Enabling Instrument," 790.

24. Brody, "Enabling Instrument," 790.

25. Morton Subotnick interview, August 19, 2015.

26. Howard Taubman, "The Arts: A Critic's View; 1958 Brussels Exposition Set a Mark That New York Is Far from Matching," *New York Times*, April 22, 1964.

27. Herbert Blau, *The Impossible Theater: A Manifesto* (Macmillan, 1964), 6.

28. Morton Subotnick interview, August 19, 2015. In this part of the interview, Subotnick was reading to me directly from a draft of his memoir, which is forthcoming from the MIT Press.

29. William Shakespeare, *King Lear* (Classic Books Company, 2001), 3.4.12–14; 183.

30. Blau, *Impossible Theater*, 285–86.

31. Blau, *Impossible Theater*, 285–86.

32. Robert LaVigne, "Robert LaVigne: A Cyberspective," Big Bridge 6, accessed March 14, 2018, https://www.bigbridge.org/issue6/rlscroll14.htm.

33. Alfred Frankenstein, "Subotnick's 'New' Music: 'An Heroic Vision' Is All of That," *San Francisco Chronicle*, September 26, 1961, 36.

34. Frankenstein, "Subotnick's New Music." Subotnick has not contributed any materials related to *Sound Blocks*—neither score nor magnetic tapes—to his publisher, nor to the tape archive at Mills College.

35. Michael McClure, "The Flowers of Politics, I," in *The New American Poetry: 1945–1960*, by Donald Allen (University of California Press, 1960), 334–53.

36. McClure, "Flowers of Politics"

37. McClure's obsession with heroism, light, and vision in this poem may have been inspiration for Subotnick's subtitle, "An Heroic Vision."

38. Morton Subotnick, "Morton Subotnick: The Mad Scientist in the Laboratory of the Ecstatic Moment," interview by Frank J. Oteri, October 1, 2013, https://nmbx.newmusicusa.org/morton-subotnick-the-mad-scientist-in-the-laboratory-of-the-ecstatic-moment/.

39. Frankenstein, "Subotnick's New Music."

40. Emphasis in original. Blau, *Impossible Theater*, 249.

41. Antonin Artaud, *The Theater and Its Double* (New York: Grove Weidenfeld, 1958), 20. The *Oxford English Dictionary* dates the first use of *black box* to describe a theatrical space to 1971; the Black Box Group, an English experimental theater company, had used the term to describe their productions since 1970. See John Epstein et al., *The Black Box: An Experiment in Visual Theatre* (Latimer New Dimensions 2, 1970).

42. Blau, *Impossible Theater*, 250.

43. Joseph Masco, "Nuclear Technoaesthetics: Sensory Politics from Trinity to the Virtual Bomb in Los Alamos," *American Ethnologist* 31, no. 3 (August 2004): 349–73.

44. Masco, "Nuclear Technoaesthetics," 4.

45. It also featured, in a later episode, a sequence by R. G. Davis, founder of the San Francisco Mime Troupe, with whom Subotnick and Sender had collaborated at the SFTMC's location at 1537 Jones Street. See *The Computer and the Mind of Man*, pt. 1, *Logic by Machine* (KQED-TV/National Educational Television and Radio Center, 1962), http://archive.org/details/ComputerAndTheMindOfMan P1LogicByMachine; *The Computer and the Mind of Man*, Program 3, "The Universal Machine" (KQED-TV/National Educational Television and Radio Center, 1962), http://archive.org/details/ComputerAndTheMindOfManP3TheUniversal Machine.

46. *The Computer and the Mind of Man*, Program 1, *Logic by Machine*.

47. Ronald R. Kline, *The Cybernetics Moment, or Why We Call Our Age the Information Age* (Johns Hopkins University Press, 2017), 4.

48. *The Computer and the Mind of Man*, Program 3, "The Universal Machine."

49. The series' producer asked Subotnick to create a score of "electronic music" to accompany the first episode; but at that point Subotnick had no access to electronic instruments that could produce sound, owning only a single-track tape recorder. Later, after he had purchased some relatively inexpensive Hewlett-Packard Oscillators, Subotnick made the soundtrack to the fourth episode of this series with "actual" electronic instruments, but the show's producers rejected them, saying that the oscillators' square waves sounded too much like clarinets. Morton Subotnick interview, August 19, 2015; Subotnick in Bernstein, *San Francisco Tape Music Center*, 119.

50. *The Computer and the Mind of Man*, Program 3, "The Universal Machine."

51. Subotnick has characterized this report as "a lecture that McLuhan had given that was going to become the new book [*Understanding Media*]." Although some of the text of the report provided by Stern would eventually make it into the 1964 publication of *Understanding Media*, there are significant differences. See Subotnick, "Morton Subotnick: Mad Scientist."

52. Gerd Stern, "From Beat Scene Poet to Psychedelic Multimedia Artist in San Francisco and Beyond, 1948–1978," interview by Victoria Morris Byerly, Oral History, 1996, Regional Oral History Office, The Bancroft Library, University of California, Berkeley, 67.

53. Subotnick, "Morton Subotnick: Mad Scientist."

54. Marshall McLuhan, *Report on Project in Understanding New Media* (National Association of Educational Broadcasters, June 30, 1960), 13.

55. McLuhan, *Report on Project in Understanding Media*, 1.

56. McLuhan, *Report on Project in Understanding Media*, 2.

57. Fred Turner, *The Democratic Surround: Multimedia and American Liberalism from World War II to the Psychedelic Sixties* (University of Chicago Press, 2015), 3, 9.

58. Allport quoted in Turner, *Democratic Surround*, 48.

59. Turner, *Democratic Surround*, 48.

60. See, for example, the first half of *Subotnick: Portrait of an Electronic Music Pioneer* (Waveshaper Media 2022).

61. An earlier incarnation of the Sound Mobile, entitled Sound Mobile without the "environmental" modifier, appeared on the program of the premiere of the SF Mime Troupe's Event II at the SFTMC in January 1963. The Mime Troupe was run by R. G. "Ronnie" Davis, who had previously been a director with the Actor's Workshop.

62. "Robert A. Dhaemers, Artist," *East Hampton Star*, April 3, 2014, http://easthamptonstar.com/Obituaries/2014403/Robert-Dhaemers-Artist.

63. The wooden supports can be seen in a very quick shot in the KQED coverage. To produce sound for several hours without changing tapes, several strategies could have been used; "performance reels" could have been made to run at very slow speeds, or there could have been some kind of automatic control mechanism that queued one machine to begin after another one ended.

64. It is impossible to tell if these sounds were recorded on site or added in later; however, some segments include the sounds of audience members reacting to the performance, which suggests that at least those were recorded on site. "Darius Milhaud Part II: Paris & California," National Educational Television: The Creative Person (KQED, 1965), San Francisco Bay Area Television Archive, https://diva.sfsu.edu/collections/sfbatv/bundles/206370.

65. "Let It Sing!," *Time* 81, no. 23 (June 7, 1963): 83. See also Turner, *Democratic Surround*, 262.

66. Milhaud Festival Program, Mills College Special Collections, CCM Archive Concert Programs, 1960–1964.

67. Milhaud Festival Program, Mills College Special Collections, CCM Archive Concert Programs, 1960–1964.

68. Subotnick, "Morton Subotnick: Mad Scientist." By late 1965, Subotnick apparently began telling other people that his work had "moved on" from Happenings. "Interview between BRC and MS, with JKS and JEB in attendance." September 29, 1965 [also dated December 30, 1965], RAC folder 3566, box 413, series 200, RG 1.2.

69. Blau, *Impossible Theater*, 250.

70. Anthony Martin interview, September 4, 2017.

71. Subotnick and Burns, Red Bull Music Academy: Morton Subotnick.

72. McLuhan, *Report on Project in Understanding New Media*, 14.

73. Of course, composers such as Babbitt saw the quantization of these parameters as a boon, rather than a limitation: because the instrument could produce thousands of possible sounds as "steady states," and because such states could change at varying speeds and could be variously combined, the total number of possible sounds that the RCA could produce was extremely large. See Babbitt, "Introduction to the R.C.A. Synthesizer."

74. Subotnick, quoted in Roads, "Interview with Morton Subotnick," 12.

75. Subotnick, "Morton Subotnick: Mad Scientist."

76. Morton Subotnick, "The Electric Heat of Creativity—Remembering Donald Buchla (1937–2016)," New Music Box, October 26, 2016, http://www.newmusicbox.org/articles/the-electric-heat-of-creativity-remembering-donald-buchla-1937-2016/. Also described in T. J. Pinch and Frank Trocco, *Analog Days: The Invention and Impact of the Moog Synthesizer* (Harvard University Press, 2002), 39; Subotnick quoted in David W. Bernstein, *The San Francisco Tape Music Center: 1960s Counterculture and the Avant-Garde* University of California Press, 2008, 114. This instrument incorporates a technology described in Alexander Rehding, "Of Sirens Old and New," in *Oxford Handbook of Mobile Music Studies* , ed. Sumanth Gopinath and Jason Stanyek (Oxford University Press, 2014), 2:77–106.

77. See, for example, Norman McClaren, *Pen Point Percussion* (National Film Board of Canada, 1951), https://www.nfb.ca/film/pen_point_percussion/. By 1964, many others had developed such "optical synthesis" techniques, including Daphne Oram's unique Oramics machine, Percy Grainger's Free Music Machine, and Yevgeny Murzin's ANS Synthesizer, among many others. It is unclear, however, if Subotnick was aware of any of these, since each was a unique creation of its inventor and not commercially available in San Francisco.

78. Another popular use of a spinning perforated disc in 1964 was the Nipkow disc, one of the fundamental technologies that enabled television, in which a beam of light would horizontally scan through holes in a spinning disc, producing electronic information when converted to signal by photocells. Subotnick's concept was also perhaps further informed by an electric organ that Sender had encountered, which used rotating optical discs with images of waveforms directly printed on them to generate sound; Sender had managed to get hold of some sample discs and was attempting to get the instrument donated to the center. This was likely a Kimball photocell organ, manufactured for a brief period in the early 1960s. A studio diagram of the SFTMC from late 1964 includes a space for a "WF [wave-form] photo-cell gen. Kimball, to be donated (hopefully). Studio Diagram [n.d.], box 11, folder 3, POP-UCSD, MSS 102, Special Collections & Archives.

79. Morton Subotnick, "Morton Subotnick on Making Your Own Creative Tools," interview by T. Cole Rachel, June 29, 2017, https://thecreativeindependent.com/people/morton-subotnick-on-making-your-own-creative-tools/.

CHAPTER 2. OPENING BUCHLA'S BOX

1. Subotnick has long maintained that he placed a classified ad in the *San Francisco Chronicle* sometime between late 1963 or mid-1964 seeking an engineer to build his music easel, and that shortly after the ad was placed, several

engineers simply showed up at the SFTMC (see, for example, Morton Subotnick, "The Electric Heat of Creativity—Remembering Donald Buchla (1937–2016)," New Music Box," October 26, 2016, http://www.newmusicbox.org/articles/the-electric-heat-of-creativity-remembering-donald-buchla-1937-2016/). However, there are several factors that cast doubt upon Subotnick's memory. The first is that it was not standard practice to include an address in a classified ad; usually only a phone number was listed, so it would be unusual for an ad respondent to show up at an address instead of calling first. The second is that, in both the *Chronicle*'s digital and microfilm archives of the years 1963 and 1964, there are no classified ads matching Subotnick's description that use either the address or phone numbers associated with the SFTMC. Members of the SFTMC had placed other classified advertisements in 1962 to sell an Ampex tape recorder, which used a telephone number associated with their first location at 1537 Jones Street (TUxedo 5-5644); the phone number associated with their second location at 321 Divisadero Street, MArket 6-0414, was listed in several *Chronicle* articles about the center throughout 1963–1965 but never used in an ad. It is possible that Subotnick placed the ad in the *Examiner*, the *Chronicle*'s competitor, but this is unlikely since the *Chronicle* had a larger circulation, and since members of the SFTMC had placed other ads in its classified section. It is also possible Subotnick placed this ad with his home phone number, but that would contradict a narrative in which respondents to the ad came directly to the SFTMC without calling first based on the address listed therein.

2. Subotnick, "Electric Heat of Creativity."

3. Curtis Roads and Morton Subotnick, "Interview with Morton Subotnick," *Computer Music Journal* 12, no. 1 (1988): 12; see also Subotnick, quoted in David W. Bernstein, *The San Francisco Tape Music Center: 1960s Counterculture and the Avant-Garde* (University of California Press, 2008), 114.

4. Marshall McLuhan, *Report on Project in Understanding New Media* (National Association of Educational Broadcasters, June 30, 1960), 29.

5. McLuhan, *Report on Project*, 1.

6. The term had been used in print first in San Francisco in 1958 by the musician and radio engineer Henry Jacobs to describe his "Vortex" performances at San Francisco's Morrison Planetarium. Henry Jacobs, "Vortex Comes Home—With Something New Added," *San Francisco Chronicle*, December 14, 1958.

7. Alden Jenks, "Alden Jenks Oral History," interview by MaryClare Bryztwa and Tessa Updike, November 11, 2013, San Francisco Conservatory of Music Library & Archives.

8. Though the University of California's physics department and the Lawrence National Laboratory were deeply involved in atomic warfare and the space race, its Physics Department did not bear this exact name. Bill Maginnis's recollection is quoted in T. J. Pinch and Frank. Trocco, *Analog Days: The Invention and Impact of the Moog Synthesizer* (Harvard University Press, 2002), 37.

9. Maginnis, quoted in in Bernstein, *San Francisco Tape Music Center*, 198.

10. "Resume—Donald F. Buchla," January 1986, Vasulka Archive, http://www.vasulka.org/archive/Artists1/Buchla,DonaldF/CV.pdf.

11. There is widespread disagreement about when Buchla delivered his system to the SFTMC. A 2008 chronology of the SFTMC by Thomas Welsh has the Buchla Box as delivered in "late 1964" (Welsh in Bernstein, *San Francisco Tape Music Center*, 276); in that same volume, Buchla remembered that he began working on it in 1963 and delivered it in 1964 (Buchla in Bernstein, *San Francisco Tape Music Center*, 169), and Subotnick remembered it being present at the premiere of Terry Riley's *In C* on November 4, 1964 (Subotnick in Bernstein, *San Francisco Tape Music Center*, 178). Despite Buchla's and Subotnick's memories, however, archival traces (or lack thereof) suggest that the first Buchla system was likely delivered to the SFTMC no earlier than late 1965. The earliest audiotapes in the SFTMC's archive collected at the Mills Center for Contemporary Music that contain recordings of the Buchla system are dated November 1965; their creator, studio technician William Maginnis, remembers making those tapes immediately after the system's delivery. Maginnis also remembers receiving the Buchla system "four to six months" after the center's brief employment of an engineer named Peter de Blanc, who was listed as a "member-engineer" of the SFTMC in a May 1965 grant application to the Rockefeller Foundation, which corroborates the late 1965 delivery date ("Proposal: Tape Music Center at Mills College," RAC, box 413, folder 3566, series 200R, RG 1.2). In addition, no mention of Buchla or his system is found in any correspondence between SFTMC managers Pauline Oliveros and Ramon Sender in 1964, and similarly no mention is found in any diary entry by grant officers of the Rockefeller Foundation, who met with Subotnick frequently throughout 1964 and 1965 and took detailed notes, until a diary entry from December 30, 1965 (Diary of Rockefeller Foundation Officer Boyd Compton, December 30, 1965, RAC, box 76, RG 12, Officers' Diaries, FA118).

12. Morton Subotnick, *Silver Apples of the Moon* (Nonesuch H-71174, 1967).

13. Such a characterization is still common: for example, the performer-composer Sarah Belle Reid used this term to describe the Buchla Modular Electronic Music System held at Mills College in 2024. "Exploring the 1st Buchla 100 Modular Synthesizer," posted February 2, 2024, https://www.youtube.com/watch?v=CiqhM-bNBrY&ab_channel=SarahBelleReid.

14. Theodore Roszak, *The Making of a Counter Culture: Reflections on the Technocratic Society and Its Youthful Opposition* (University of California Press, 1995), 42.

15. On September 7, 1965, a front page, below-the-fold newspaper article in the *San Francisco Examiner* described Haight-Ashbury as full of "long-haired hippies," the first recorded use of the phrase; it also prominently noted the "bewildering, though democratic, assemblage of homosexuals and lesbians" at

Romeo's, a nearby bar on Haight street. Michael Fallon, "Bohemia's New Haven," *San Francisco Examiner*, September 7, 1965.

16. Fred Turner, *From Counterculture to Cyberculture: Stewart Brand, the Whole Earth Network, and the Rise of Digital Utopianism* (University of Chicago Press, 2010), 38.

17. "Resume—Donald F. Buchla." The extent of Buchla's involvement with NASA is difficult to verify, but Buchla recalled two specific projects in an interview with the author on February 5, 2016, both dealing with biometric telemetry.

18. In one experiment, Buchla implanted electroencephalogram (EEG) sensors directly into the skulls of a breed of large, black rabbits from Texas, which garnered criticism from those opposed to animal testing. But as Buchla slyly remembered, he did not think these rabbits were treated cruelly because they spent most of their time during the experiment copulating. Donald Buchla interview, February 6, 2016.

19. Manfred E. Clynes and Nathan S. Kline, "Cyborgs and Space," *Astronautics*, September 1960, 27.

20. Peter Galison, "The Ontology of the Enemy: Norbert Wiener and the Cybernetic Vision," *Critical Inquiry* 21, no. 1 (October 1994): 233, https://doi.org/10.1086/448747.

21. Norbert Wiener, *Cybernetics: Or Control and Communication in the Animal and the Machine*, 2nd ed. (MIT Press, 1961), 42.

22. Stefan Helmreich, "Potential Energy and the Body Electric: Cardiac Waves, Brain Waves, and the Making of Quantities into Qualities," *Current Anthropology* 54, no. S7 (October 1, 2013): 139–48.

23. "Resume—Donald F. Buchla."

24. Buchla recounts that the LEDs he used came directly from RCA and cost "$450 each" (equivalent to roughly $4,500 in 2025). Buchla, quoted in Bernstein, *San Francisco Tape Music Center*, 170.

25. Buchla, quoted in Bernstein, *San Francisco Tape Music Center*, 170.

26. As Mara Mills has argued, "the rhetorics of charity, rehabilitation, and cybernetic enhancement have marginalized and even erased the specificity of disability in most historical accounts." See Mara Mills, "On Disability and Cybernetics: Helen Keller, Norbert Wiener, and the Hearing Glove," *Differences* 22, no. 2-3 (January 1, 2011): 90; Buchla, quoted in Bernstein, *San Francisco Tape Music Center*, 170.

27. Buchla, quoted in Bernstein, *San Francisco Tape Music Center*, 164.

28. Ortiz, "People Electronic Music 3/14/64," BANC PIC 2006.029-NEG, box 1427, sleeve 138806.04, frame 3B, Pictorial Collection, The Bancroft Library, University of California.

29. Buchla, quoted in Bernstein, *San Francisco Tape Music Center*, 164.

30. Morton Subotnick, "Morton Subotnick: The Mad Scientist in the Laboratory of the Ecstatic Moment," interview by Frank J. Oteri, October 1, 2013;

Morton Subotnick, "Morton Subotnick on Making Your Own Creative Tools," interview by T. Cole Rachel, June 29, 2017.

31. Letter from Ramon Sender to Vladimir Ussachevsky, October 17, 1962, CPEMCA. Studio technician William Maginnis remembers attempting to fix this machine in early 1964 and having the tape heads fall to pieces in his hands, because they were not laminated together. William Maginnis and Ramon Sender interview, July 5, 2015.

32. William Maginnis, "Interview by Eric Chasalow and Barbara Cassidy, video, June 27, 1997, The Video Archive of Electronic Music, http://ericchasalow.com/oralhist.html.

33. Maginnis and Sender interview, July 5, 2015. Although this device was never operational, Subotnick apparently mentioned it as an example of the SFTMC's technological innovations in discussions with Boyd Compton, a grant officer for the Rockefeller Foundation, with whom he began meeting in May 1965. RAC, box 413, folder 3655, 200R, San Francisco Tape Music Center.

34. One of Maginnis's other self-described kludges was an attempt to filter out a 60Hz hum in the studio with an assortment of unlabeled toroidal coils scrapped from teletype machines, which Subotnick colloquially called "soap chips." Maginnis, "Interview by Eric Chasalow and Barbara Cassidy."

35. Pressing a key would light an incandescent bulb, causing the resistance of the CdS cell to fall and producing a smooth "envelope" that could control the level of any sound source. This envelope could be further controlled by adjusting the current being sent to the incandescent bulb, which Maginnis remembers discovering serendipitously. Luc Secretan, "Cadmium-Sulfide Cell Micrometeorite Detector," National Aeronautics and Space Administration, 1962,, NASA Space Science Data Coordinated Archive, https://nssdc.gsfc.nasa.gov/nmc/experiment Display.do?id=1962-070A-05.

36. Maginnis, "Interview by Eric Chasalow and Barbara Cassidy."

37. Buchla, quoted in Bernstein, *San Francisco Tape Music Center*, 170.

38. Anthony Martin interview, September 7, 2017.

39. Subotnick, quoted in Bernstein, *San Francisco Tape Music Center*, 166.

40. Stephen Heinemann, "Pitch-Class Set Multiplication in Theory and Practice," *Music Theory Spectrum* 20, no. 1 (1998): 72.

41. Subotnick, quoted in Bernstein, *San Francisco Tape Music Center*, 166.

42. Subotnick, quoted in Bernstein, *San Francisco Tape Music Center*, 166.

43. Sender and Maginnis interview, July 1, 2015. In Subotnick's memory, initial conversations with Buchla took place between 1963 and 1964, with the instrument arriving at some point in 1964. Subotnick has also long maintained that he asked for and received $500 from the Rockefeller Foundation to give to Buchla to produce the system. However, no documentation of this transaction exists in the detailed daily diary entries kept by RF grant officer Boyd Compton, who first met Subotnick in April 1965 and subsequently documented twelve

additional meetings with him between May and December 1965. It is in theory possible that $500 fell below some kind of internal limit for discretionary spending, but Compton's meticulous and exhaustive diary practice would suggest that this was not the case. The RF did award the SFTMC a grant in June 1965 to fund its 1965–1966 season; however, the proposed budget prepared by Sender and Subotnick does not specify any line item for the Buchla Box. It is possible that Subotnick simply gave Buchla $500 from these funds. See Diary of Rockefeller Foundation Officer Boyd Compton, December 30, 1965, RAC, box 76, RG 12, Officers' Diaries, FA118.

44. See Subotnick, quoted in Bernstein, *San Francisco Tape Music Center*, 115. As instrument builder Mark Verbos has noted, Buchla used "CTS potentiometers C&K toggle switches, Switchcraft audio jacks and Johnson banana jacks," all mounted by hand. Buchla also used some of the earliest available light-emitting diodes (LEDs), which were extremely rare and expensive. Mark Verbos, "Made from the Best Stuff on Earth," *Buchla Tech* (blog), July 24, 2014, http://buchlatech.blogspot.com/2014/07/made-from-best-stuff-on-earth.html.

45. Subotnick, "Electric Heat of Creativity."

46. Separating these types of electronic signal also had technical benefits: audio signal operated at a low 1 volt RMS, since it needed to have headroom for momentary transient peaks; control signal was a steady-state signal of 0 to +15 volts; pulse signal was a momentary +15 volt pulse wave. This allowed for more prevision in control signal while maintaining a clean audio signal without distortion.

47. Harald Bode, "A New Tool for the Exploration of Unknown Electronic Music Instrument Performances," *Journal of the Audio Engineering Society* 9, no. 4 (October 1961): 264.

48. Bode, "New Tool," 265.

49. Many twentieth-century electronic musical instruments such as the Trautonium and others had already included extensive use of "formant filters" to synthesize sounds that replicated the spectral features of human vocal production. See Thomas Patteson, "'A New, Perfect Musical Instrument:' The Trautonium and Electric Music in the 1930s," in *Instruments for New Music: Sound, Technology, and Modernism* (University of California Press, 2015).

50. Roland Wittje has described this functional equivalency as an act of translation that produced abstractly identical representations of different systems. See Roland Wittje, "The Electrical Imagination: Sound Analogies, Equivalent Circuits, and the Rise of Electroacoustics , 1863–1939," in *Music, Sound, and the Laboratory from 1750–1980* (University of Chicago Press, 2013), 44.

51. Many remember that Buchla built modules with whatever parts were accessible in his shop, instead of following a standardized schematic. Fellow instrument designers jokingly described this as "selling prototypes"; William Maginnis remembers an early module that simply consisted of tack-soldered components, without even a circuit board. Roger Linn interview, February 4, 2016.

52. Pinch and Trocco, *Analog Days*, 18.

53. Trevor Pinch, "Why You Go to a Piano Store to Buy a Synthesizer," in *Path Dependence and the Social Construction of Technology*, ed. R. Garud and P. Karnoe (LEA Press, 2001), 391. A later published version of this interview in Pinch and Trocco's *Analog Days* differs in a significant way: it quotes Buchla as saying "a keyboard is dictatorial" (44). This has been widely requoted in dozens of journalistic sources in the wake of Buchla's 2016 death, including his *New York Times* obituary. Jon Pareles, "Don Buchla, Inventor, Composer and Electronic Music Maverick, Dies at 79," *New York Times*, September 17, 2016. However, Buchla designed many instruments with black-and-white keyboards throughout his career.

54. Buchla also designed a third Touch Controlled Voltage Source for David Tudor, the Model 113, which contained four circular pads with labyrinthine patterns to control voltage in two simultaneous dimensions. Donald F. Buchla, *The Modular Electronic Music System* (Buchla & Associates, 1966).

55. Donald Buchla interview, February 6, 2016. For more on the history of deafness, cybernetics, and manual/digital interfaces, see Mills, "On Disability and Cybernetics."

56. Donald F. Buchla, interview by Torsten Schmidt, Red Bull Music Academy, 2007, http://www.redbullmusicacademy.com/lectures/don-buchla-passing-the-acid-test.

57. "Tuning" the keyboard would require a frequency counter, oscilloscope, and/or strobe tuner, not included on Buchla's instrument but standard instruments for electronic music studios.

58. Buchla's touch-sensitive metal plates are often described as working with the skin's capacitance. However, as Chip Flynn and Mark Milanovich have shown, some early touch-controlled instruments, such as the Model 113 built for David Tudor and housed at Wesleyan University, were actually resistive, not capacitive; this meant that the metal touch plates held a +15 volt charge that was partially absorbed by the human touching the instrument. The MEMS Project, "Touchplates," accessed February 28, 2024, https://www.memsproject.info/touchplates.

59. Sender and Maginnis interview, July 1, 2015.

60. Sender had modified the SFTMC's machine to play tape loops outside of the cartridges by exposing its playback heads, turning the unit on its side and manually tensioning tape loops on the playback heads with gravity. Sender used this instrument, and an earlier technological solution for controlling its playback heads, to create a tape work variously titled *World Food III* (CCM CD 123, Center for Contemporary Music Archive, Special Collections, F. W. Olin Library, Mills College) or *Worldfood VII* (Locust Music L55, 2006).

61. Buchla makes these interconnections between modules clear in his 1966 brochure; though he suggests that the Model 107 be "normally" controlled by the

Model 114, he also suggests that the Model 114 can control gates, mixers, and attack waveform generators. Buchla, *Modular Electronic Music System*.

62. Although the minimum and maximum values of 5 Hz and 20 kHz were specified on these oscillator's front panels, the calibration of each individual oscillator varied significantly. On Model 144 and 158 oscillators installed in an early system delivered to the CPEMC, for example, studio users used a frequency counter to manually determine these values and wrote them in grease pencil on the panels. As of 2025, these markings were still visible.

63. Buchla, *Modular Electronic Music System*.

64. Joel Chadabe, *Electric Sound: The Past and Promise of Electronic Music* (Prentice Hall, 1997), 147.

65. Sender and Maginnis interview, July 1, 2015.

66. CCM CD 24, Center for Contemporary Music Archive, Special Collections, F. W. Olin Library, Mills College.

67. Subotnick, "Electric Heat of Creativity."

68. Turner, *From Counterculture to Cyberculture*, 38.

69. Bob Gluck, "Nurturing Young Composers: Morton Subotnick's Late-1960s Studio in New York City," *Computer Music Journal* 36, no. 1 (2012): 67.

70. The etymology of this word derived from the Greek *psychế* (mind) and *dêlos* (visible/manifest); it was coined by the psychiatrist Humphry Osmond in 1957, during correspondence with the author Aldous Huxley. See Janice Hopkins Tanne, "Humphry Osmond," *BMJ: British Medical Journal* 328, no. 7441 (March 20, 2004): 713.

71. Emphasis in original. Richard Alper et al.,"Rationale of the Mexican Psychedelic Training Center," in *Utopiates: The Use & Users of LSD 25*, edited by Richard Blum (Atherton Press, 1964), 179.

72. A 2011 article by Robert Greenfield estimates that Stanley brought "3,600 capsules" of LSD to the Bay Area, but since Stanley was fond of extremely large "heroic" doses—up to 200 micrograms per capsule—the quantity of normal doses in Stanley's original batch might have been significantly higher. See Robert Greenfield, "Owsley Stanley: The King of LSD," *Rolling Stone*, March 14, 2011, https://www.rollingstone.com/feature/owsley-stanley-the-king-of-lsd-82181/.

73. Oliveros did participate in an evening of exploring peyote with Ramon Sender, along with Oliveros's partner Laurel Johnson, in the summer of 1963. Pauline Oliveros to Terry Riley, August 28, 1963, POP-NYPL, box 23, folder 4. But Sender remembers Oliveros becoming frightened by Johnson's "bad trip" and deciding not to pursue the drug further. See Ramon Sender, Ramon Sender Oral History, interview by Mary Clare Bryztwa and Tessa Updike, April 14, 2014, San Francisco Conservatory of Music Library & Archives.

74. Peter Richardson, *No Simple Highway: A Cultural History of the Grateful Dead* (St. Martin's Press, 2015), 52.

75. Jill D'Alessandro and Colleen Terry, *Summer of Love: Art, Fashion, and Rock and Roll* (University of California Press, 2017), 23.

76. De Blanc was listed as a "member engineer" on the SFTMC's grant application to the Rockefeller Foundation, dated May 20, 1965. "Proposal: Tape Music Center at Mills College," RAC, box 413, folder 3566, series 200R, RG 1.2.

77. Sender and Maginnis interview, July 1, 2015. In a 2008 reminiscence of his time at the SFTMC, Maginnis generously pseudonymized de Blanc as "Joe McGee;" after realizing that de Blanc had been lying about his credentials as an engineer, Maginnis wrote, "Well, it turned out that Mr. McGee's degree was not from McGill but from crystal meth, good old crank strikes again!" Maginnis, in Bernstein, *San Francisco Tape Music Center*, 201.

78. Ramon Sender to Boyd Compton, September 6, 1965, RAC, box 413, folder 3566, series 200, RG 1.2.

79. "Subotnick Bound for New York," *San Francisco Examiner*, January 11, 1966, declares that his move was "imminent." RAC, box 413, folder 3566, 200R. Subotnick has said several times that he took thirteen months to create *Silver Apples*, which was released in July 1967; this implies that he began work on the album in June 1966. The studio he set up at 144 Bleecker Street had been purchased by NYU in January 1966, and in this period, he also utilized some of the taped material from his Buchla to create *Prelude No. 4*. Email correspondence with Robert Gluck, March 11, 2024. Boyd Compton interview with Robert Corrigan, January 10, 1966, Boyd Compton diaries, RAC, box 75, FA118, RG 12.

80. The Family Dog began as a communal house at 2125 Pine Street, called the "Dog House"; the Family Dog began when members of this house teamed up with the inhabitants of nearby 1836 Pine Street, which housed Red Dog Saloon visualist Bill Ham, as well as Chet Helms and Janis Joplin. Helms would later manage Big Brother and the Holding Company, which eventually featured Joplin as their singer. See Joel Selvin, *San Francisco: The Musical History Tour; A Guide to Over 200 of the Bay Area's Most Memorable Music Sites* (Chronicle Books, 1996), 200.

81. The first concert, titled "A Tribute to Dr. Strange" and held on October 16, 1965, was the first of three "tributes" to comic book characters produced by Helms. Each "tribute" concert featured Helms's band, the Charlatans, as well as other bands such as Jefferson Airplane and the Great Society.

82. Sender and Maginnis interview, July 1, 2015. In advertisements placed in the January 21, 22, and 23, 1965, issues of the *San Francisco Chronicle*, the SFTMC's involvement was listed as "Ramon Sender & The Tape Center Group," as well as "Don Buchla's Machine."

83. Interview with Ramon Sender, San Francisco Conservatory of Music Library & Archives, and Oral History Project, 2016. .

84. Babbs: "He set up his synthesizer scene on the balcony next to our electronic setup. He had ten speakers spread out on the balcony which ran on three

sides of the hall. He had a keyboard where you could run your fingers along it and have the sound zoom around through the speakers. He let me run his stuff whenever I wanted." Ken Babbs, Sky Pilot home page, accessed October 26, 2015, https://www.skypilotclub.com/index.1.html.

85. Tom Wolfe, *The Electric Kool-Aid Acid Test* (Bantam Books, 1999), 381.

86. Lou Gottlieb, "The World of Music: Glorious Electricity," *San Francisco Chronicle*, January 18, 1966.

87. "Electronic Music Program," *San Francisco Chronicle*, January 27, 1966.

88. Carl Cunningham, "Light and Sound in Berkeley." *San Francisco Chronicle*, January 29, 1966.

89. Although sometimes erroneously attributed to Buchla, *Furthur*'s original time lag system was engineered by Sandy Lehmann-Haupt, one of the original Pranksters who had been working professionally as a sound engineer when the bus was first purchased and modified in 1964. See Wolfe, *Electric Kool-Aid Acid Test*, 68, for a description of Lehmann-Haupt's initial construction of the lag; hanging microphones are seen in various shots of the bus from 1964 included in the 2011 film *Magic Bus*, which comprised a long-awaited edit of footage shot by Kesey on the bus's 1964 trip.

90. Barry Miles, *Hippie* (Sterling Publishing, 2005), 32; Pinch and Trocco, *Analog Days*, 93.

91. These two donations represented a significant financial contribution to the AAA. 100,000μg of LSD was, in the words of Kampman, "enough for about 500 hefty wallops. Nowadays more like a thousand" (Lars Kampman, "Lars' Memoirs," Pittendrigh dis org, accessed April 22, 2018, http://pittendrigh.org, 15); several AAA members remembered selling some of the LSD to fund the eventual purchase of land in Colorado for a permanent home for the AAA. A brochure produced in April 1966 by Buchla lists the price of a two-cabinet system such as the one purchased for the AAA at $3,420, which is the equivalent of roughly USD$34,000 in 2025.

92. Kampman remembers that Subotnick "had dibs on the first one. Then we got the second one. There was no prototype and Len, who was to play the synthesizer in the group, spent much time in Oakland at Buchla's studio, determining what the instrument should do." Kampman, "Lars' Memoirs," 24.

93. Leary et al., "Rationale of the Mexican Psychedelic Training Center," 176–86.

94. Richard Alpert and Sydney Cohen, *LSD* (New American Library, 1966), 15.

95. Alpert and Cohen, *LSD*, 15.

96. Kampman's memoirs describe an emotionally intimate relationship with Alpert, although both men were still closeted at the time. Kampman, "Lars' Memoirs."

97. Kampman remembers this event as "August 1966," but Ken Babbs remembers it as July 24, 1966.

98. Lee Quarnstrom, *When I Was a Dynamiter! Or, How a Nice Catholic Boy Became a Merry Prankster, a Pornographer, and a Bridegroom Seven Times*, ed. Iris Berry (Punk Hostage Press, 2014), 203.

99. Kampman, "Lars' Memoirs," 20.

100. Richard P. McKeon et al., eds., *On Knowing—The Social Sciences* (University of Chicago Press, 2017), xix.

101. McKeon et al., *On Knowing*, xix, 3.

102. McKeon et al., *On Knowing*, 2.

103. Frazer devoted himself to the Buchla Box throughout the 1960s and into the 1970s, bringing it with him on a 1969 tour across the country, and eventually to the site of the AAA's international community in the Huerfano Valley in southern Colorado; although the AAA mostly disintegrated after Kampan's death from AIDS in the 1980s, the instrument has featured prominently in the surviving community's remembrances of the commune, including several photographs posted on social media sites in the 2000s.

104. *Psych-Out* (American International Pictures, 1968).

105. The AAA did leave behind visual artifacts, such as psychedelic concert posters; the group's bassist, Trixie Merkin, famously posed topless for a 1969 *Rolling Stone* article about "groupies." The group was also documented in a 1970 film by Gordon Quinn, which contained footage of a live performance at the University of Chicago in 1968. See Jerry Hopkins et al., "The Groupies and Other Girls," *Rolling Stone*, February 15, 1969; and *Anonymous Artists of America* (Kartemquin Films, 1970).

106. Juliette Goodrich and Molly McCrea, "Repair of Iconic '60s Era Synthesizer Turns into Long, Strange Trip for Engineer." CBS News Bay Area, 2019, https://www.cbsnews.com/sanfrancisco/news/1960s-rock-music-san-francisco-lsd-buchla-100-synthesizer-grateful-dead/. Stanley ordered his system in 1966 but never took possession of it because he fled the country when the manufacture and possession of LSD became increasingly criminalized. Red panel modules from this system were sold off individually, ending up in various other systems, including one at Cal State Hayward (now Cal State East Bay), where Curtis had his trip, as well as one at the CPEMC. One such red panel module, a Module 111 Dual Ring Modulator, is held in Vladimir Ussachevsky's papers at the Library of Congress; discoloration on the front panel suggests that he mostly used it as a coaster. My own experiences licking this module, as well as a red panel Model 144 Dual Square Wave Generator now housed in the system at the Columbia Computer Music Center, produced no psychedelic effects.

107. Hunter S. Thompson, *Hell's Angels: A Strange and Terrible Saga* (Ballantine Books, 1996), 227–28. This party was also the subject of Allen Ginsberg's "First Party at Ken Kesey's with Hell's Angels," in *Planet News: 1961–1967* (City Lights Publishers, 2001).

108. Thompson, *Hell's Angels*, 228.

109. Suzanne Ciani, "Don Buchla Memorial by Suzanne Ciani," *Huffington Post* (blog), October 6, 2016, https://www.huffingtonpost.com/suzanne-ciani/-don-buchla-memorial-by-s_b_12378852.html.

110. Buchla, quoted in Bernstein, *San Francisco Tape Music Center*, 172.

111. Buchla, *Modular Electronic Music System*.

112. Suzanne Ciani, present during this interview, quickly countered Subotnick's rhetorical question by answering: "I'm the sucker that ended up playing it." Morton Subotnick and Suzanne Ciani, "Music Pioneers: Suzanne Ciani and Morton Subotnick in Conversation," interview by Frosty, Red Bull Music Academy, June 24, 2016, http://daily.redbullmusicacademy.com/2016/06/encounters-suzanne-ciani-morton-subotnick.

113. This term was a popular metaphor for psychedelics such as peyote, possibly originating in Carlos Castaneda's bestselling 1968 pseudo-ethnography, *The Teachings of Don Juan: A Yaqui Way of Knowledge*.

114. Donald Buchla and Charles MacDermed, "The Electric Music Box," *Synthesis*, January 1971, box 32, Robert Moog papers, no. 8629, Division of Rare and Manuscript Collections, Cornell University Library, Ithaca, NY.

115. Pauline Oliveros, "Improvising Composition: How to Listen to the Time Between," in *Negotiated Moments: Improvisation, Sound, and Subjectivity*, ed. Gillian H. Siddall and Ellen Waterman, Improvisation, Community, and Social Practice (Duke University Press, 2016).

CHAPTER 3. THE PATCHWORK GIRL

1. Oliveros told this anecdote several times later in her life, though it is impossible to verify if these two events occurred on the same day. Indeed, as discussed in chapter 2, this chronology clashes with Maginnis's memory of the Buchla Box being delivered in late 1965, since Oliveros's early tape works were produced in early to mid-1965. See Pauline Oliveros, "Reverberations: Eight Decades," *Jefferson Journal of Science and Culture*, no. 2 (2012): 45.

2. Dempster, quoted in David W. Bernstein, *The San Francisco Tape Music Center: 1960s Counterculture and the Avant-Garde* (University of California Press, 2008), 263.

3. Pauline Oliveros, "Improvising Composition: How to Listen to the Time Between," in *Negotiated Moments: Improvisation, Sound, and Subjectivity*, ed. Gillian H. Siddall and Ellen Waterman, Improvisation, Community, and Social Practice (Duke University Press, 2016).

4. Emphasis added. Pauline Oliveros to David Tudor, April 4, 1965, DTP, box 57, folder 8.

5. L. Frank Baum, *The Patchwork Girl of Oz* (Reilly & Lee, 1913).

6. Pauline Oliveros, *Sounding the Margins: Collected Writings 1992–2009* (Deep Listening Publications, 2010), 28.

7. John Cage, *Silence: Lectures and Writings* (Wesleyan University Press, 1973), 51. See also Douglas Kahn, "John Cage: Silence and Silencing," *Musical Quarterly* 81, no. 4 (1997): 556–98, http://www.jstor.org/stable/742286.

8. E. T. A. Hoffmann, "Beethoven's Instrumental Music: Translated from E. T. A. Hoffmann's 'Kreisleriana' with an Introductory Note," trans. Arthur Ware Locke, *Musical Quarterly* 3, no. 1 (1917): 123–33.

9. Erickson's textbook cites both systems theorist Paul Weiss and biologist Wladimir Weidlé to justify this assertion. Robert Erickson, *Sound Structure in Music* (University of California Press, 1975), 82.

10. Riley, quoted in Bernstein, *San Francisco Tape Music Center*, 207.

11. Pauline Oliveros, interview by Hannah Bächer, Red Bull Music Academy, 2016, https://www.redbullmusicacademy.com/lectures/pauline-oliveros-lecture.

12. All five tapes of these improvisations are archived on CCM CD 150, Center for Contemporary Music Archive, Special Collections, F. W. Olin Library, Mills College; they have also been uploaded to the Internet Archive by Other Minds: *Early Improvisations with Pauline Oliveros, Loren Rush, Terry Riley, and Others*, 1957, http://archive.org/details/C_1957_XX_XX.

13. Oliveros, "Improvising Composition."

14. Pauline Oliveros, "Tripping on Wires: The Wireless Body; Who Is Improvising?," *Critical Studies in Improvisation/Études Critiques En Improvisation* 1, no. 1 (2004), http://www.criticalimprov.com/article/view/9/31.

15. Emphasis in original. This quote was later collected by Erickson's colleague Roger Reynolds from three sources: "Chapter 10 of an unpublished autobiography, the transcript of an extraordinarily candid and revealing UCSD Music Department Seminar that occurred in 1983, and an interview by one of his occasional instrumentalist collaborators, Daniel Dunbar." "Erickson Celebration at UCSD," POP-NYPL, box 7, folder 15.

16. Erickson: "I don't think that plain piano technique can be developed thru improvisation. Separate or extra problem. However, musicianship can be developed as you well know, and musical things, are the most important." Emphasis in original. Robert Erickson to Pauline Oliveros, October 2, 1960, POP-NYPL, box 7, folder 14.

17. This gynocentric group suggests a much earlier starting point to the kind of sociality that Oliveros would later develop in the ♀ Ensemble in San Diego and suggests that Oliveros was cultivating queer spaces well before her public identification as a lesbian in 1971. See Martha Mockus, *Sounding Out: Pauline Oliveros and Lesbian Musicality* (Routledge, 2008), 76.

18. Emphasis transcribed from tape. "Oliveros Interview with Glenn Glasow, KPFA Radio, 7/21/1960," CCM CD 189.

19. Ramon Sender, "Ambulancia en el desierto: Ramón Sender y el San Francisco Tape Music Center," interview by Rubén Coll, September 9, 2016, Radio del Museo Reina Sofia, http://radio.museoreinasofia.es/ambulancia-desierto.

20. The visual accompaniment was attributed to Milton Cohen, known for his *Space Theatre* installations in Ann Arbor in the 1960s and his involvement with the ONCE Group. According to Gordon Mumma, Cohen had been "developing a theater based on projected images [since 1956 in San Francisco]" and had become a professor at the University of Michigan in Ann Arbor in 1957. Mumma recalls that he and Robert Ashley formed the Cooperative Studio for Electronic Music in 1958 "at first largely to supply electronic music for the Space Theatre." Gordon Mumma and Michelle Fillion, *Cybersonic Arts: Adventures in American New Music*, Music in American Life (University of Illinois Press, 2015), 8–10.

21. Fred Frith, "The 1958 KPFA Tapes," in Bernstein, *San Francisco Tape Music Center*, 210.

22. Mildred Schroeder, "[. . .] to Beethoven?," *San Francisco Examiner*, December 1961, POP-NYPL, box 25, folder 14.

23. Ramon Sender remembers this as a "bring your own loudspeaker" concert, in which the audience was invited to bring whatever speakers they had on hand, due to the financial limitations of the Conservatory. Ramon Sender interview, July 1, 2015.

24. POP-UCSD, box 13, folder 24.

25. Milton Cohen, "Space Theatre," *Dimension*, no. 14 (1963): 62, quoted in Emily Weingarten, *The Music of ONCE: Perpetual Innovation*, Lulu.com, 2008, 3.

26. Pauline Oliveros, liner notes for *Four Electronic Pieces 1959–1966*, SR185 (Sub Rosa, 2008), https://www.subrosa.net/en/catalogue/early-electronic-music/pauline-oliveros.html.

27. Heidi von Gunden argues that the vocal sounds in *Sound Patterns* explicitly mimic the sounds of these electronic instruments. I would go one step further to argue that Oliveros's work renders the human body and the instrument of the electronic studio functionally identical: each is part of a total instrumental system that yields novel results from exploratory improvisation. Heidi Von Gunden, *The Music of Pauline Oliveros* (Scarecrow Press, 1983), 30.

28. Letters addressed to Oliveros during this time indicate her address as the Gaudeamus Festspielhaus in Bilthoven. POP-NYPL, box 25, folder 16.

29. David Tudor, "Major Figures in American Music: David Tudor," 241 a-l, interview by Jack Vees, August 1995, Yale Oral History of American Music, Yale University Libraries.

30. Pauline Oliveros to David Tudor, ca. October 1963, DTP, box 57, folder 8; David Tudor to Pauline Oliveros, March 12, 1964, POP-NYPL, box 27, folder 27.

31. Pauline Oliveros to David Tudor, ca. October 1963, DTP, box 57, folder 8. Oliveros also remembered this in her tribute to Tudor after his passing in the

mid-1990s. Pauline Oliveros, "DAVID TUDOR: Testimonials," August 19, 1996, http://davidtudor.org/Life/testimonials.html.

32. Aldous Huxley, *Island* (Perennial, 2009).

33. Correspondence between Oliveros and Harris suggests that the two also had a romantic relationship, though it is unclear when it began. POP-NYPL, Music Division, box 24, folder 19.

34. You Nakai, "On the Instrumental Natures of David Tudor's Music" (PhD diss., New York University, 2016), 143n45.

35. The title also included the line: "staging conceived and directed by Elizabeth Harris." "Three Concerts with David Tudor," POP-UCSD, box 13, folder 27.

36. Photographs of the event were taken for the *San Francisco Examiner* and are held in the Bancroft Library at UC Berkeley. Bryant, "Musicians—Oliveros, Tudor, Sender," BANC PIC 2006.029-NEG, box 1427, Sleeve 138825.04, Frame E7, Pictorial Collection, The Bancroft Library, University of California.

37. Oliveros, "Improvising Composition." Different versions of this anecdote also appear in Oliveros, "Reverberations: Eight Decades" and Pauline Oliveros, "What Matters? Make the Music!," Contact!, September 2015, http://econtact.ca/17_3/oliveros_music.html.

38. Pauline Oliveros, interview by Hannah Bächer, 2016.

39. That is, if Oliveros played a low A 2 with a fundamental frequency of 110 Hz and a low A# 2 with a fundamental frequency of 116.5 Hz, she would perceive an acoustic "beat' with a frequency of 6.5Hz

40. Oliveros, "What Matters?"

41. This may have been due to the fact that the massive, floor-to-ceiling patch bay was built to military specifications, supposedly sourced from a B-12 bomber. See William Maginnis, interview by Eric Chasalow and Barbara Cassidy, June 27, 1997, The Video Archive of Electronic Music, http://ericchasalow.com/oralhist.html.

42. Oliveros, "What Matters?"

43. On the Ampex 350 tape machine, which had a distance of ~2 inches between record and play heads, the delay time was 266 milliseconds if the tape ran at 7.5 ips (inches per second); if it ran at the more standard 15 ips, the delay time would be only 133 milliseconds. See Pauline Oliveros, "Tape Delay Techniques for Electronic Music Composers," *Composer* 1, no. 3 (December 1969).

44. Pauline Oliveros to David Tudor, January 20, 1966, DTP, box 57, folder 8.

45. Maggi Payne has been tirelessly digitizing Oliveros's massive tape archive since the early 1990s and has grouped these two untitled tracks along with the other extant *Mnemonics* tapes onto CCM CD 152.

46. This effect of mixing oscillator signals with tape bias signals was quickly observed and exploited by many other members of the SFTMC, particularly Bill Maginnis. Maginnis and Sender interview, July 5, 2015.

47. Though Oliveros used the Buchla in other ways, producing the *Bog* series of tapes in the late 1960s at Mills College, she ultimately remembered it as

qualitatively different from her system at the SFTMC. "In 1966, the solid-state Buchla modular synthesizer rendered the classical electronic music studio obsolescent. I made some real-time pieces performing the Buchla with my delay system, but missed the sound of the tube oscillators. Transistor oscillators have a very different sound quality and feeling." Oliveros, *Sounding the Margins*, 255.

48. Some tapes were given uniquely punny titles, for example, *Big Mother Is Watching You, No Mo, Participle Dangling*; others have more descriptive titles, such as *Fed Back at 2:40AM* and *The Day I Disconnected the Erase Head and Forgot to Reconnect It*. Most of these tapes have been collected and archived at the Mills College Center for Contemporary Music; many were also reissued in 2012 by Important Records. Pauline Oliveros, *Reverberations: Tape & Electronic Music 1961–1970* (Important Records IMPREC352, 2012).

49. Chase occupied a large warehouse at 136 Embarcadero. Other performers that day included Peter Cohen of the S.F. Mime Troupe, the composer Robert Moran, Allie Hilder, Constance Beeson, and Orville Dale. "Tape-Athon: Electronic music by Pauline Oliveros, July 22, 1967," POP-UCSD, box 13, folder 30.

50. Only one tape from this period saw a contemporaneous commercial release on an LP record: the tape *I of IV*, produced at the UTEMS studio in 1966, which was included on the composer David Behrman's record *New Sounds in Experimental Music* on Columbia Odyssey in 1967 alongside works by Steve Reich and the late Richard Maxfield. This would be the only tape issued on a commercially available LP until the inclusion of *Bye Bye Butterfly* on a compilation album released by 1750 Arch Records in 1977, which initiated a period of critical acclaim for that work. Steve Reich, Richard Maxfield, and Pauline Oliveros, *New Sounds in Electronic Music (Come Out/Night Music/I of IV)*, Music of Our Time (Odyssey 32 16 0160, 1967).

51. Kerry O'Brien, "Listening as Activism: The 'Sonic Meditations' of Pauline Oliveros," *New Yorker*, December 9, 2016, https://www.newyorker.com/culture/culture-desk/listening-as-activism-the-sonic-meditations-of-pauline-oliveros.

52. "The ♀ Ensemble Brochure," POP-UCSD, box 12, folder 1; Pauline Oliveros, "Sonic Meditations," in *Source*, ed. Larry Austin, Douglas Kahn, and Nilendra Gurusinghe, Music of the Avant-Garde, 1966–1973 (University of California Press, 2011), 342–46, http://www.jstor.org.remote.baruch.cuny.edu/stable/10.1525/j.ctt1png6w.73.

53. Mockus, *Sounding Out*, 2.

54. Pauline Oliveros to David Tudor, January 20, 1966, DTP, box 57, folder 8; various artists, *New Music for Electronic and Recorded Media* (1750 Arch Records S-1765, 1977). In 1980, the *New York Times* critic John Rockwell singled this recording out as "one of the most beautiful pieces of electronic music to emerge from the 60's." John Rockwell, "The Musical Meditations of Pauline Oliveros," *New York Times*, May 25, 1980.

55. Mockus, *Sounding Out*, 23.

56. Von Gunden, *Music of Pauline Oliveros*, 55, 57.

57. Mockus connects the butterfly to genital anatomy through the novelist Monique Wittig's 1986 novel *The Lesbian Body*, which she quotes at length. Mockus, *Sounding Out*, 27.

58. The studio director in this anecdote was most certainly Hugh LeCaine, likely afraid that Oliveros's experiments would damage the instruments in the studio— which, unlike the military-surplus kludges at the SFTMC, were purpose-built instruments with less room for electrical failure. Pauline Oliveros, "Some Sound Observations," in *Software for People: Collected Writings 1963–1980* (Smith Publications; Printed Editions, 1984), 17–27, 27.

59. Oliveros, "Sonic Meditations."

60. Douglas Kahn, *Earth Sound Earth Signal: Energies and Earth Magnitude in the Arts* (University of California Press, 2013), 174.

61. Oliveros, "Sonic Meditations."

62. This is Lucier's paraphrase from a 1981 interview with William Duckworth. Alvin Lucier, *Reflections: Interviews, Scores, Writings/Reflexionen: Interviews, Notationen, Texte* Edition MusikTexte 003 (MusikTexte, 1995), 322.

63. Pauline Oliveros, "Quantum Improvisation: The Cybernetic Presence," keynote address presented at the conference Improvisation Across Borders at UCSD, April 11, 1999, https://www.hz-journal.org/n16/oliveros.html.

CHAPTER 4. ALVIN LUCIER AND THE AMBIGUITY OF SOUND AND SIGNAL

1. Harry James Sands Jr., "Panmunjom Today," *Sylvania Scanner*, May 1967, ALP, box 1, folder 1.

2. Peter Galison, "The Ontology of the Enemy: Norbert Wiener and the Cybernetic Vision," *Critical Inquiry* 21, no. 1 (October 1994): 223.

3. Galison, "Ontology of the Enemy," 223.

4. ALP, box 1, folder 1.

5. ALP, box 1, folder 1.

6. ALP, box 1, folder 1.

7. Edmond M. Dewan, "Occipital Alpha Rhythm Eye Position and Lens Accommodation," *Nature* 214, no. 5092 (June 1967): 975–77, https://doi.org/10.1038/214975a0.

8. "Eeek / Lunk / Eeoww / Tump / Rrrooomm / Oooop / Noozle / Lingg / Clack / Mooooo / Ugh / Shoosh / Izzle / Clump," *Sylvania Scanner*, May 1967, ALP, box 1, folder 1.

9. *No Ideas but in Things: The Composer Alvin Lucier* (Wergo, 2013).

10. Christoph Cox, "The Alien Voice: Alvin Lucier's North American Time Capsule 1967," in *Mainframe Experimentalism: Early Computing and the Foundations of the Digital Arts*, ed. Hannah Higgins and Douglas Kahn (University of California Press, 2012), 178.

11. Cox, "Alien Voice," 178. Cox's invocation of the Lacanian concept of the "real" differs from most other Lacanian scholarship, which holds that the real is always absolutely imperceptible, rather than perceptible. See Bruce Fink, *The Lacanian Subject: Between Language and Jouissance* (Princeton Univ. Press, 1997).

12. Cox, "Alien Voice," 183.

13. Christoph Cox, "Beyond Representation and Signification: Toward a Sonic Materialism," *Journal of Visual Culture* 10, no. 2 (August 1, 2011): 157. More recently, the German musicologist Bernhard Rietbrock has argued for an alternate usage of the Lacanian concept of the "real" to understand Lucier's works, differentiating between a "pre-symbolic and a symbolically immanent real." Bernhard Rietbrock, *Alvin Lucier's Reflexive Experimental Aesthetics* (Wolke Verlag, 2022), 42.

14. Christoph Cox, *Sonic Flux: Sound, Art, and Metaphysics* (University of Chicago Press, 2018), 103.

15. Volker Straebel and Wilm Thoben, "Alvin Lucier's Music for Solo Performer: Experimental Music beyond Sonification," in *Collected Work: Organised Sound* (Cambridge University Press, 2014), 17–29.27.

16. Straebel and Thoben, "Alvin Lucier's Music for Solo Performer," 27.

17. Kahn, *Earth Sound Earth Signal*, 101.

18. Alvin Lucier, *Chambers: Scores by Alvin Lucier* (Wesleyan University Press, 1980).

19. For more on "diffraction," see Karen Barad, "Diffracting Diffraction: Cutting Together-Apart," *Parallax* 20, no. 3 (July 3, 2014): 168–87; Donna Haraway, "The Promises of Monsters: A Regenerative Politics for Inappropriate/d Others," in *Cultural Studies*, ed. Lawrence Grossberg, Cary Nelson, and Paula A. Treichler (Routledge, 1992), 295–337; Trinh T. Minh-Ha, "She, the Inappropriate/d Other: Introduction," *Discourse* 8 (Fall–Winter 1986/1987): 3–10.

20. Lucier, *Chambers*, 70.

21. Lucier et al., *Reflections*, 28; ALP, box 12, folder 3.

22. ALP, box 12, folder 3.

23. ALP, box 12, folders 2–4; box 13, folders 1–3; box 55, folders 12–13; box 56, folders 1–2.

24. Kahn, *Earth Sound Earth Signal*, 95.

25. Douglas Kahn, "The Military-Arts Nexus: Two Cases in the United States, c. 1970," *Studies in Material Thinking* 8 (May 2012): 4.

26. Dewan, "Occipital Alpha Rhythm Eye Position," 976.

27. G dome, "Brainwave Control Device by Edmond Dewan, 1964," posted July 10. 2013, https://www.youtube.com/watch?v=FGXMLuDVz_Q.

28. Straebel and Thoben have documented multiple sources of this anecdote and include one such 2001 retelling in Straebel and Thoben, "Alvin Lucier's Music for Solo Performer," 29.

29. Alvin Lucier, "Notes in the Margins," in *Reflections*, 510, quoted in Kahn, *Earth Sound Earth Signal*, 90.

30. Kahn, *Earth Sound Earth Signal*, 86.
31. Kahn, *Earth Sound Earth Signal*, 86.
32. Andrew Pickering, *The Cybernetic Brain: Sketches from Another Future* (University of Chicago Press, 2010), 22.
33. Pickering, *Cybernetic Brain*, 85.
34. Kahn, *Earth Sound Earth Signal*, 88.
35. Roger T. Dean, *The Oxford Handbook of Computer Music* (Oxford University Press, USA, 2009), 247.
36. Emphasis added. Andrew Raffo Dewar, "Reframing Sounds: Recontextualization as Compositional Process in the Work of Alvin Lucier," *Leonardo Music Journal* 22 (2012): 5.
37. Lucier, *Chambers*, 71.
38. Melissa M. Littlefield, *Instrumental Intimacy* (Johns Hopkins University Press, 2018), 4.
39. Littlefield, *Instrumental Intimacy*, 9. For more on the political consequences of such instrumental intimacy, see Danielle Carr, "Political Anatomies of the Cyborg: Liberal Subjects and Neural Engineering" (PhD diss., Columbia University, 2023), https://doi.org/10.7916/bsfr-y255.
40. Cox, *Sonic Flux*, 103.
41. Cox, "Alien Voice," 178.
42. For a detailed accounting of the histories of these materials, as well as a thorough bibliography of interviews about the work's performance, see Straebel and Thoben, "Alvin Lucier's Music for Solo Performer," 18n2–3.
43. ALP, box 41, folder 4.
44. ALP, box 41, folder 4.
45. ALP, box 41, folder 4. On a CV from late 1966, Lucier lists the work as "Composition for Amplified Lip," and lists its premiere by the Sonic Arts Group, with Robert Ashley on "Solo Lip," at the Rose Art Museum, Brandeis University, Waltham, Massachusetts, April 22, 1966.
46. ALP, box 41, folder 4.
47. ALP, box 41, folder 4.
48. ALP, box 41, folder 4.
49. Letter from Alvin Lucier to Gordon Mumma, April 4, 1966, ALP, box 1, folder 11.
50. Straebel and Thoben, "Alvin Lucier's Music for Solo Performer," 27.
51. Letter from Alvin Lucier to T. C. Brownfield, November 11, 1966, ALP, box 3, folder 3.
52. See, for example, Lucier's comments on *Vespers* in *Chambers*, 15–27.
53. Lucier, *Chambers*, 19.
54. These pages are in ALP, box 56, folder 3, which is labeled "Synthesizer Patch Settings 1972." Curiously, they are not collected in ALP, box 53, folder 7, which contains materials related to *Vespers*. This may have been a cataloging

error; the ink on these pages and their paper is similar to other notebooks Lucier used in 1967 and which relate to *Vespers.*

55. John Durham Peters, "'Memorable Equinox': John Lilly, Dolphin Vocals, and the Tape Medium," *Boundary 2* 47, no. 4 (2020): 1-2. See also D. Graham Burnett, "Adult Swim: How John C. Lilly Got Groovy (and Took the Dolphin with Him), 1958–1968," in *Groovy Science: Knowledge, Innovation, and American Counterculture,* ed. David Kaiser and W. Patrick McCray (University of Chicago Press, 2016), 18.

56. John C. Lilly, *The Mind of the Dolphin* (Doubleday, 1967), 99.

57. Lilly, *Mind of the Dolphin,* 99.

58. Lilly, *Mind of the Dolphin,* 93.

59. Lilly, *Mind of the Dolphin,* 93.

60. ALP, box 56, folder 3

61. ALP, box 56, folder 3.

62. Lilly, *Mind of the Dolphin,* 95–96.

63. Lilly, *Mind of the Dolphin,* 95.

64. Norbert Wiener, *Cybernetics: Or Control and Communication in the Animal and the Machine* (MIT Press, 1948), x.

65. ALP, box 53, folder 7.

66. ALP, box 53, folder 7.

67. The proposal is loosely formatted like a grant proposal, including subheadings such as "Problems and Objectives," "Relevant Literature," and "Equipment Needed." In the equipment section, Lucier lists an "Environ-Ears Binaural recording system, available on order from Listening, Inc." This suggests that Lucier wrote this equipment list after encountering the Sondol, which was also manufactured by Listening, Inc. However, the multiple drafts of the proposal, written in different ink and on different pages of the same notebook, do not provide a definitive timeline of Lucier's development of this proposal vis-a-vis his encounter with Listening, Inc.

68. ALP, box 56, folder 6. These quotations are from a draft version of the essay "The Tools of My Trade," eventually published in Thomas Delio, *Contiguous Lines: Issues and Ideas in the Music of the '60's and '70's* (University Press of America, 1984).

69. Batteau is listed as Clear No. 533 in "Now There Are 606 Clears," *The Auditor: The Journal of Scientology* 30 (1967); Dwight W. Batteau, "How to Make a Robot Speak English," *Analog Science Fiction and Fact,* August 1964. Listening, Inc. may have also been a North American producer of "E-Meters" for the Church of Scientology after the ban on their importation from England by the US Food and Drug Administration in the early 1960s. Batteau served as the "auditor" to his employee, Stephen Moshier, who had dropped out of Harvard to become a Scientologist and develop the "Man-to-Dolphin Translator." In 1962 Ross Moshier, Stephen Moshier's father and a research chemist at the Air Force

Aeronautical Research Laboratory in Ohio, wrote letters to the American Medical Association, the Bureau of Customs, the Department of Health, Education and Welfare, and the Food and Drug Administration, complaining that Batteau had built an "E-Meter" and had "audited" Moshier for twenty-five hours and then billed him for $550. See Ross Moshier to Oliver Field, September 4, 1962, https://scientology-research.org/documents/fda-documents/, https://archive.org/details/fda.files/01.%20Folder1/page/n479/mode/2up.

70. William Moeser, "Whiz Kid, Hands Down," *Life*, September 14, 1962. Flanagan would continue to develop para- and pseudoscientific products throughout his long life, including pyramidical devices related to his idiosyncratic theory of "pyramid power," along with various amulets and nutritional supplements sold through his company PhiSciences and advertised on YouTube. After his passing in 2019, his disciples established the Patrick Flanagan Library in Sedona, Arizona.

71. Dwight W. Batteau and Peter R. Markey, *Man/Dolphin Communication*, Final Report, US Naval Ordinance Test Station, China Lake, California, December 15, 1966–December 13, 1967.

72. Batteau and Markey, *Man/Dolphin Communication*, 6–7. In a 1967 article in *Electronic Design*, Batteau's research program using these instruments was contrasted with Lilly's approach at the CRI, with the author concluding that Batteau was doing little more than "conditioning [dolphins] to exhibit a type of behavior." Richard Einhorn, "Dolphins Challenge the Designer," *Electronic Design*, December 6, 1967, 63.

73. Listening, Inc., "Annual Report," *Analog Science Fiction and Fact*, July 1967.

74. Orthography in original. Listening, Inc., "Annual Report," 54.

75. Lucier, *Chambers*, 21–22, 124.

76. Dating *Vespers* is difficult, since Lucier did not produce a score until well after the work had been performed several times. Notebooks and sketches such as those found in ALP, box 12, folder 3, and box 38, folder 16 are dated both 1967 and 1968, but later publications cite the work as originating in 1969.

77. The Sondol's patent, filed in 1968 and issued in 1971, also lists William Van Lennep and Edmund Perry as coinventors. Van Lennep had apparently graduated from Harvard College in 1929; was briefly married to Rebecca Pollard Van Lennep Guggenheim Logan, a philanthropist and socialite; and served as the curator of the Harvard Theatre Collection from 1940 to 1960 (in addition to sharing a name with a noted nineteenth-century homeopathist). Little is known about Perry except that he was a CB radio operator who lived in Cambridge, Massachusetts. It is perhaps not surprising that Moshier, as a Scientologist, would be interested in developing electronic technologies to effect physical (i.e., sonic and electrical) cybernetic control of the human brain; as Scientology was intimately connected to the antipsychiatric movement in the 1960s, it promoted

a psychophysiological, rather than a psychoanalytic or psychopharmacological, approach to human consciousness. See Danielle Carr, "Scientology's Lonely Turf War," Pioneer Works, April 20, 2021, https://pioneerworks.org/broadcast/scientology-psychiatry.

78. Stephen Moshier, William Van Lennep, and Edmund Perry, Acoustic Locator, 3558822 (Cambridge, MA, filed May 22, 1968, and issued January 26, 1971). Although there is some evidence that the Sondol was advertised nationally, Lucier seems to have been its main user in the 1960s and beyond. The Sondol is listed, among several other devices intended to aid the blind, in a supplement to the 1968 *Research Bulletin of the American Federation for the Blind*, but its commercial availability seems to have ended by the 1970s.

79. Indeed, Lilly spends an entire chapter in *The Mind of the Dolphin* (61–99) describing differences in human and cetacean auditory apparatuses, although an extremely complex technological attempt is made to allow each to communicate with the other.

80. Bruce Clarke, "John Lilly, *The Mind of the Dolphin*, and Communication Out of Bounds," *Communication +1* 3, no. 8 (2014): 10.

81. ALP, box 38, folder 16.

82. ALP, box 38, folder 16. Although Listening, Inc.'s literature describes the Sondol as a portmanteau of "sonic dolphin," Lucier's sketches and eventual score use the term "sonar dolphin," perhaps due to the Sondol's functional similarity to sonar ("sound navigation and ranging") technologies.

83. ALP, box 38, folder 16.

84. Lilly, *Mind of the Dolphin*, 91.

85. ALP, box 36, folder 16.

86. Cox, *Sonic Flux*, 103.

87. In these notes, Lucier even proposed an alternate title to the work: "Vespers (1968): Acoustic Orientation by Means of Echolocation with Visual Analog." ALP, box 53, folder 7.

88. ALP, box 12, folder 4. Although *Listening in the Dark* was included in the bibliography for the research project proposal discussed in this chapter, Lucier's archive only contains cursory notes on a few chapters. Though Lucier would later cite Griffin's work as the original inspiration for *Vespers*, his archival notes from 1967 suggest it was not as initially influential as Lilly's work was.

89. ALP, box 12, folder 4.

90. ALP, box 12, folder 4.

91. The film footage from this shoot was used in two separate films: *Wasserpfeifen in New-York: Musikalische Avantgarde zwischen Ideologie und Elektronik* (Water pipes/bongs in New York: Musical Avant-garde between ideology and electronics), dir. Hans G. Helms, 1972, and *New Music: Sounds and Voices from the Avant-Garde New York 1971*, dir. Michael Blackwood, 1972.

92. ALP, box 12, folder 4.

93. ALP, box 12, folder 2.

94. ALP, box 12, folder 2.

95. Alvin Lucier, "Alvin Lucier—Thoughts on Installations," Kunstradio—Radiokunst Online, accessed August 7, 2024, https://www.kunstradio.at/ZEITGLEICH/CATALOG/ENGLISH/lucier-e.html.

96. Daniel James Wolf, "No Ideas But in Things—The Composer Alvin Lucier—Text by Daniel James Wolf," 2013, http://alvin-lucier-film.com/text_wolf.html.

97. Wolf, "No Ideas But in Things."

98. Cox, *Sonic Flux*, 13.

99. Pickering, *Cybernetic Brain*, 21.

100. Pickering, *Cybernetic Brain*, 22.

101. Rietbrock, *Alvin Lucier's Reflexive Experimental Aesthetics*, 141.

102. William Carlos Williams, "Paterson," in *The Collected Earlier Poems of William Carlos Williams* (New Directions Books, 1938), 231–235. https://www.fadedpage.com/showbook.php?pid=20200926.

103. ALP, box 12, folder 3.

104. Intergalactic Music: Rare Sun Ra, "Sun Ra 1971 Oakland, CA Pre European Tour Rehearsal" (WKCR, May 22, 1995), posted April 6, 2022, https://www.youtube.com/watch?v=QLnp_1cLNlA.

CHAPTER 5. SUN RA AND THE MINIMOOG

1. Richard Williams, "Sun Ra Coming!", *Melody Maker*, October 10, 1970, AACSR Box 3, Folder 4.

2. For a narrative of Schoenfeld's involvement with this tour, see Benjamin Piekut, "Indeterminacy, Free Improvisation, and the Mixed Avant-Garde: Experimental Music in London, 1965–1975," *Journal of the American Musicological Society* 67, no. 3 (December 1, 2014): 769–824, https://doi.org/10.1525/jams.2014.67.3.769.

3. Although Ra was a prolific recording artist, his music was only commercially available by mail order through his own labels, notably Saturn Records, run by his spiritual and business associate Alton Abraham in Chicago. Since he had never performed in Europe, it is highly likely that the only LPs circulating in Europe had been obtained by critics such as Williams, who were dedicated to keeping abreast of American experimental and avant-garde music.

4. Richard Williams, "Night of the Giants," *Melody Maker*, November 7, 1970, AACSR, box 3, folder 4.

5. For an exhaustive list of reviews of this concert see Piekut, "Indeterminacy, Free Improvisation," 794 n18.

6. Ronald Atkins, "Sun Ra," *Guardian*, November 10, 1970, AACSR, box 3, folder 4.

7. Richard Williams, "Sun Ra, Queen Elizabeth Hall," *The Times*, November 10, 1970, AACSR, box 3, folder 4.

8. For a narrative description of this tour, see John Szwed, *Space Is the Place: The Lives and Times of Sun Ra* (Duke University Press, 1997), 280–85. Szwed characterizes the West African percussionists Math Samba, from Guinea, and Roger Aralamon Hazoumé, from the Republic of Dahomey (now Benin), as "African dancers," and claims that Hazoumé was also a fire-eater; recordings from this tour list Samba as playing percussion and Hazoumé as playing the balafon, as well as dancing. For more on Sun Ra's complicated relationship to modern Africa, see Tam Fiofori, "Sun Ra & Africa," *BookArtVille* (blog), May 23, 2020, https://bookartville.com/sun-ra-africa/.

9. As quoted in Robert Greenfield, "Sun Ra & Europe's Space Music Scene," *Rolling Stone*, January 21, 1971. In 1970, Greenfield was employed at *Rolling Stone*'s London bureau, so it is likely that he attended the November 9 performance at Queen Elizabeth Hall.

10. According to Greenfield, at the Donaueschingen Festival the German crowd booed the Arkestra; Ra responded by proclaiming that "booing is the sound of sub-humans and I'm playing for the angels," which shocked the postwar German audience into silence.

11. AACSR, box 15, folder 4.

12. For a list of instruments used on this tour see Szwed, *Space Is the Place*, 280. The concert program for the November 9, 1970, performance, held in AACSR, box 15, folder 4, suggests that the Farfisa and Hohner instruments were lent to Ra for this tour with financial help of the US embassy in London. The "Spacemaster" organ was described by *DownBeat* journalist J. C. Thomas on June 13, 1968 ("Sun Ra's Space Probe") as "an organ especially manufactured for him by the Chicago Musical Instrument Co. that sounds like a cross between a theremin and bagpipes." Szwed comes perilously close to plagiarizing this exact same description in *Space Is the Place*, 226. During this period, Ra performed with a Kalamazoo K-101/Gibson G-101 transistorized frequency divider combo organ (featured, for example, on 1967's *Atlantis*), which was capable of producing loud, droning timbres that fit Thomas's description. Both the Kalamazoo and Gibson brands were owned by CMI, and so it is likely that the "Spacemaster" was simply Ra's Kalamazoo K-101/Gibson G-101.

13. Intergalactic Music: Rare Sun Ra, "Sun Ra 1971 Oakland, CA Pre European Tour Rehearsal" (WKCR, May 22, 1995), posted April 6, 2022, https://www.youtube.com/watch?v=QLnp_1cLNlA.

14. For Wiener, a "black box" is a device that "performs a definite operation on the present and past of the input potential, but for which we do not necessarily have any information of the structure by which this operation is performed. On the other hand, a white box will be a similar network in which we have built the relation between input and output potentials in accordance with a definite structural plan for securing a previously determined input-output relation." Norbert Wiener, *Cybernetics: Or Control and Communication in the Animal and the Machine* (MIT Press, 1948), xiii1.

15. Louis Onuorah Chude-Sokei, *The Sound of Culture: Diaspora and Black Technopoetics* (Wesleyan University Press, 2016), 85.

16. Chude-Sokei, *Sound of Culture*, 82.

17. CAA, SR-R073, reel 46.

18. Tam Fiofori, "The Illusion of Sun Ra," *Liberator* 7, no. 12 (December 1967): 13–15, AACSR, box 1, folder 5.

19. Amiri Baraka, "Jazzmen: Diz & Sun Ra," *African American Review* 29, no. 2 (1995): 249–55, https://doi.org/10.2307/3042302, 254.

20. Erik Steinskog, *Afrofuturism and Black Sound Studies: Culture, Technology, and Things to Come*, Palgrave Studies in Sound (Springer International/Palgrave Macmillan, 2018), 126, https://doi.org/10.1007/978-3-319-66041-7.

21. Kodwo Eshun, *More Brilliant Than the Sun: Adventures in Sonic Fiction* (Quartet Books, 1998), 160.

22. *Space Is the Place*, 1974.

23. Frank Adams and Burgin Mathews, *Doc: The Story of a Birmingham Jazz Man* (University of Alabama Press, 2012), 79.

24. Burgin Mathews, "Sun Ra in Birmingham: A Few Ear(th)ly Artifacts," *Burgin Mathews* (blog), May 22, 2020, https://burginmathews.com/2020/05/22/sun-ra-in-birmingham-a-few-earthly-artifacts/.

25. Ra's personal use of early sound recording technologies has also been celebrated as evidence of his interest in technology. John Szwed even goes so far as to claim that Ra was using a steel audiotape recording device manufactured by the Brush Development Company called a Soundmirror as early as 1937 (Szwed, *Space Is the Place*, 34). Szwed's claims are dubious, however; the first commercially available home tape recorder made by the Brush Development Company was introduced at some point in the late 1940s, and sources differ as to the exact year: either 1945, in David Morton, *Sound Recording: The Life Story of a Technology* (JHU Press, 2006), 211, or 1947, in Eric D. Daniel et al., *Magnetic Recording: The First 100 Years* (John Wiley & Sons, 1998), 77.

26. According to Ra discographer Robert Campbell, this recording was made on July 29, 1948, at 5414 South Prairie Avenue; other sources report it was made in 1953. It was eventually released on Saturn Records (Saturn 485) as "Dreams Come True" (or alternately, "Deep Purple"). Robert Campbell et al., "From Sonny Blount to Sun Ra: The Chicago Years," 2021, revised May 11, 2025, http://campber.people.clemson.edu/sunra.html.

27. The Tornados, *Telstar* (Decca 45-F 11494, 1962).

28. A manuscript score for *The Magic City* is held in AACSR, box 7, folder 7.

29. Andy Beta, "Sun Ra and His Arkestra: *The Magic City/My Brother the Wind* Vol. 1," Pitchfork, October 16, 2017, https://pitchfork.com/reviews/albums/sun-ra-and-his-arkestra-the-magic-city-my-brother-the-wind-vol-1/.

30. Irwin Chusid, "Liner Notes to *Atlantis*, by Sun Ra & His Astro Infinity Arkestra," Enterplanetary Koncepts, 2014, https://sunramusic.bandcamp.com/album/atlantis.

31. A score for this composition, held in the AACSR at the University of Chicago Library, contains no indications for these timbral changes: only the pitch of the instrument is specified. AACSR, box 8, folder 1.

32. Szwed, *Space Is the Place*, 248.

33. T. J. Pinch and Frank Trocco, *Analog Days: The Invention and Impact of the Moog Synthesizer* (Harvard University Press, 2002), 29.

34. R. A. Moog Co., "Electronic Music Composition-Performance Equipment Short Form Catalog 1967," 1967, 1, https://moogfoundation.org/wp-content/uploads/1967-R.A.-Moog-Catalog.pdf.

35. R. A. Moog Co., "Electronic Music Composition-Performance Equipment," 5.

36. Thom Holmes, "Moog: A History in Recordings," June 17, 2013, https://moogfoundation.org/moog-a-history-in-recordings-by-thom-holmes-part-two/; Pinch and Trocco, *Analog Days*, 112.

37. Donal Henahan, "Is Everybody Going to the Moog?," *New York Times*, August 24, 1969.

38. As the *System* article informed readers, "Bob is the son of George Moog, an assistant general superintendent at the Astoria Transformer shop.... When Bob was younger, father and son made radios, sound mixes, amplifiers and other equipment in the basement of their Flushing home." RMP, box 32.

39. "Young People's Concert: 'Bach Transmogrified,'" CBS, broadcast April 27, 1969, posted 26, 2018, https://youtu.be/J8sS5NkADBE?si=84SXPg2eA_4RbVBD.

40. Kingsley, along with his partner Jean-Jacques Perrey, had been performing with an Ondioline—an instrument related to Ra's Clavioline—since 1965; Kingsley programmed the Moog on his 1967 album with Jean-Jacques Perrey, *Kaleidoscopic Variations*. In 1969, Kingsley released *Music to Moog By*, his first solo record with the Moog, which featured the hit single "Pop Corn."

41. Ben Young, "Liner Notes to *My Brother the Wind* Vol. I" (Cosmic Myth Records, CMR002 2017), 12. This recording session was financed by Thomas "T. S." Mims, a Chicago-based tenor saxophonist who had been involved with Thmei Research, Ra's Chicago-based occult research group with his business partner Alton Abraham. It was initially released as Sun Ra and His Astro-Solar Infinity Arkestra, *My Brother the Wind* (Saturn Records 521, 1970) and subsequently re-released with additional tracks as *My Brother the Wind Volume II* (El Saturn Records 523/SRA 2000, 1971).

42. Kingsley quoted in Irwin Chusid, "Liner Notes to *My Brother the Wind* Vol. I," emphasis added.

43. A January 1969 press release held in Moog's archive (RMP, box 47, folder 7) provides an exhaustive list of every Moog owner up to that point. These included seventy-eight colleges and universities, three scientific institutions, and eighty-nine commercial music studios or independent musicians.

44. Nelson had been introduced to the Moog in 1968 by Quincy Jones, who invited him to the residence of Paul Beaver, Moog's West Coast sales representative. Harvey Siders, "Oliver's New Twist," *DownBeat*, July 23, 1970, 17. Jones had used

Beaver's Moog system in 1967 to produce the music for the NBC crime drama *Ironside*, a later recording of which was used in Quentin Tarantino's film *Kill Bill: Volume 1* (2003). Bley's recollection is recounted in Paul Bley and David Lee, *Stopping Time: Paul Bley and the Transformation of Jazz* (Véhicule Press, 1999), 108–9. Scholars such as Bob Gluck have cast doubt upon Bley's memory, given that a complete Moog modular system would have cost in the tens of thousands of dollars. Bob Gluck, "Paul Bley and Live Synthesizer Performance," *Jazz Perspectives* 7, no. 3 (December 2013): 306–7n19, 20, https://doi.org/10.1080/17494060.2014.912257.

45. Tam Fiofori, "Myth, Music, and Media," *Glendora Review: African Quarterly on the Arts* 3, nos. 3 & 4 (2004): 113.

46. Fiofori, "Myth, Music, and Media," 113.

47. Moog's archive does include a copy of Ra's 1970 LP *My Brother the Wind*, which contained recordings from the 1969 sessions Ra conducted at Kingsley's studio, but it does not include the 1971 LP *My Brother the Wind Volume II*, which contained recordings made at Moog's studio, nor any other Sun Ra recording. A handwritten receipt held by Roger Luther, an independent historian who maintains his own collection of Moog-related material, describes the instrument as a "'Model B' Mini Moog Synthesizer," serial number 1001, and dates the transaction on July 29, 1970.

48. Moog, quoted in John Hinds, "Robert Moog Conversation with John Hinds—September 9, 1991—Stanford, Ca.," Sun Ra Research, October 2000, 42.

49. Weiss, quoted in Pinch and Trocco, *Analog Days*, 72.

50. Brian Kehew, "Liner Notes to *My Brother the Wind* Vol. I."

51. Pinch and Trocco, *Analog Days*, 217–218, 221.

52. Pinch and Trocco, *Analog Days*, 200.

53. RMP, box 47, folders 9 & 11. A business plan included with a bank loan application submitted by Moog on June 2, 1970—a few weeks before Ra's visit—makes absolutely no mention of the Minimoog. Instead, it focuses on the profitability of large, expensive studio systems and an "organ-synthesizer for home use" for "serious home amateurs, status symbol collectors, and solo pro musicians."

54. RMP, box 47, folder 9.

55. Pinch and Trocco, *Analog Days*, 222.

56. Fiofori, "Myth, Music, and Media," 113; capitalization in original. The instrument offered to Ra was not designed by Robert Moog, nor was it the first prototype.

57. Moog, quoted in Hinds, "Robert Moog Conversation with John Hinds," 42.

58. Recordings from this session were commercially released as tracks 7–11 on *My Brother the Wind Volume II*.

59. Intergalactic Music: Rare Sun Ra, "Sun Ra 1971 Oakland, CA."

60. CAA, SR-R073, reel 46.

61. AACSR, box 3, folder 3.

62. Szwed, *Space Is the Place*, 279 (emphasis added).

63. Although complete audio documentation from these three concerts is not publicly available, two LP records that contain recordings from this three-night stint are extant and bear the titles *Nuits de la Fondation Maeght*, Vols. 1 and 2. Both circulated as bootlegs for many decades before being digitally re-released by Sun Ra LLC, an organization managed by Irwin Chusid and comprised of Sun Ra's lawful heirs and rightsholders.

64. Pinch and Trocco, *Analog Days*, 223.

65. Sun Ra and His Intergalactic Research Arkestra, *It's After the End of the World* (MPS Records CRM 748, 1970) contains recordings from the Donaueschingen Festival on October 17, 1970, and the Berlin Jazz Festival on November 7, 1970; Sun Ra and the Intergalactic Research Arkestra, *Live in London* (Transparency 0317, 2005) contains a recording from November 9, 1970, at Queen Elizabeth Hall.

66. Daniel Kreiss, "Performing the Past to Claim the Future: Sun Ra and the Afro-Future Underground, 1954–1968," *African American Review* 45, nos. 1–2 (2012): 197–203, https://doi.org/10.1353/afa.2012.0006.

67. Sun Ra, *The Immeasurable Equation: The Collected Poetry and Prose*, ed. James L. Wolf and Hartmut Geerken (Waitawhile, 2005), 351.

68. Ra, *Immeasurable Equation*, 216.

69. AACSR, box 15, folder 4.

70. Tam Fiofori, "Sun Ra's African Roots," *Melody Maker*, February 12, 1972, 32.

71. Fiofori, "Sun Ra's African Roots," 32.

72. Fiofori, "Sun Ra'a African Roots," 32.

73. Szwed, *Space Is the Place*, 97–98, 281.

74. For a thorough discussion of Ra's break from the Black Panther Party, see Kreiss, "Performing the Past."

75. Szwed, *Space Is the Place*, 285.

76. Jayna Brown, *Black Utopias: Speculative Life and the Music of Other Worlds* (Duke University Press, 2021), 168.

77. Ra, quoted in Szwed, *Space Is the Place*, 115 (emphasis added).

78. Kreiss, "Performing the Past," 66.

79. AACSR, box 15, folder 4 (emphasis added).

80. Fiofori, "Sun Ra's African Roots," 32.

81. These appellations come from Ra's poems collected in Ra, *Immeasurable Equation*, respectively: "The Universe Sent Me to Converse with You" (1972/2005, 403); "Intergalactic Master" (1972/2005, 216); "Other Gods I Have Heard Of" (1980/2005, 284); and "The Other Otherness," version 2 (1980/2005, 286).

82. Ra, *Immeasurable Equation*, 216.

83. Jacson, quoted in Szwed, *Space Is the Place*, 112.

84. During a 1971 rehearsal, Ra extended this capability to other unnamed electronic instruments, reinscribing the instruments' ability to perform precisely notated music: "Now they got a lot of different instruments you don't know

nothing about. And that's what I've been trying to tell Black musicians.... I met a fella who's got a book on it this big and it takes you six months to learn how to play it because it's got the music, a blue C, green C, a red C, a white C, a purple C, and every C's get a different something on that instrument. You read the music, you see a purple C and you push this over there, or a red C and you push this over there. Now the red C might just be a little bit sharper. And in between the C and C sharp you might have eight notes or 16 notes and all of them different." Intergalactic Music: Rare Sun Ra, "Sun Ra 1971 Oakland, CA."

85. Ra, *Immeasurable Equation*, 467.
86. AACSR, box 15, folder 10.
87. Ra, *Immeasurable Equation*, 467.
88. Ra, *Immeasurable Equation*, 467–68.
89. Ra, *Immeasurable Equation*, 469.
90. Jacson, in Szwed, *Space Is the Place*, 112; emphasis in original.
91. See Sylvia Wynter, "Unsettling the Coloniality of Being/Power/Truth/Freedom: Towards the Human, After Man, Its Overrepresentation—An Argument," *CR: The New Centennial Review* 3, no. 3 (2003): 257–337. Katherine McKittrick has explored Wynter's interest in the possibilities of science, particularly the scientific concept of *autopoiesis* developed by Maturana and Varela in 1980, in "Axis, Bold as Love: On Sylvia Wynter, Jimi Hendrix, and the Promise of Science," in *Sylvia Wynter*, ed. Katherine McKittrick (Duke University Press, 2015), 142–63, https://doi.org/10.1215/9780822375852-006.
92. Orlando Patterson, *Slavery and Social Death: A Comparative Study* (Harvard University Press, 1985); Frank B. Wilderson III, *Afropessimism* (Liveright Publishing, 2021), 41–42.
93. Henry Dumas, "An Interview with Sun Ra," *Hiram Poetry Review*, no. 3 (Fall-Winter 1967): 31–32, AACSR, box 1, folder 4.
94. Édouard Glissant, *Poetics of Relation*, trans. Betsy Wing (University of Michigan Press, 1997), 6.
95. Glissant, *Poetics of Relation*, 7.
96. Glissant, *Poetics of Relation*, 111.
97. Glissant, *Poetics of Relation*, 191–92.
98. John E. Drabinski, *Glissant and the Middle Passage: Philosophy, Beginning, Abyss* (University of Minnesota Press, 2019), 140, https://doi.org/10.5749/j.ctvh4zj15.
99. Édouard Glissant, *Carribean Discourse: Selected Essays* (University of Virginia, 1999), 123–24, quoted in Fred Moten, *In The Break: The Aesthetics of the Black Radical Tradition* (University of Minnesota Press, 2003), 7.
100. Moten, *In the Break*, 7.
101. Moten, *In the Break*, 22.
102. This prose poem, undated and filed in its own folder in AACSR, box 11, folder 6, is found on the top half of a photocopy of a typewritten page; the bottom

half appears to be a program note for "New York Town," a solo piano suite that Ra premiered in New York in 1988. Separating these two halves, with irregular spacing that suggests the halves may have been copied and pasted separately, is this statement, in italics: *"By the way I could sue a lot of people who have falsely stated I was born Herman Blount."* It is unclear when the top half, or this single italicized sentence, were written.

103. Ra, *Immeasurable Equation*, 403.

104. Chude-Sokei, *Sound of Culture*, 82.

105. Kara Keeling, *Queer Times, Black Futures* (New York University Press, 2019), 66.

106. Keeling, *Queer Times, Black Futures*, 69.

107. Saidiya V. Hartman, *Scenes of Subjection: Terror, Slavery, and Self-Making in Nineteenth-Century America*, Race and American Culture (Oxford University Press, 1997), 4.

108. Hartman, *Scenes of Subjection*, 35.

109. Hartman, *Scenes of Subjection*, 36.

110. In an interview filmed in the ancient Egyptian collection held at the Philadelphia Museum of Art for the 1980 documentary *Sun Ra: A Joyful Noise*, Ra makes this connection explicit: "Somehow, ancient Egypt has been thought of as the kingdom of bondage. It would be better to say the kingdom of discipline: the kingdom of precision, the kingdom of culture, beauty, art. It would be better to say that, because this is the proof of it. The stones speak. The stones are speaking through vibrations of beauty, vibrations of discipline, vibrations of precision. Yes, the stones speak to the people of planet Earth. If teenagers are lost, it is because they have been fed upon the word freedom, not discipline." *Sun Ra: A Joyful Noise* (Robert Mugge, 1980).

111. Thom Holmes, "Sun Ra & the Minimoog," *The Bob Moog Foundation* (blog), November 6, 2013, https://moogfoundation.org/sun-ra-the-minimoog-by-historian-thom-holmes/; Paul Youngquist, *A Pure Solar World: Sun Ra and the Birth of Afrofuturism*, Discovering America (University of Texas Press, 2016), 215–16.

112. Ra, *Immeasurable Equation*, 470.

AFTERWORD

1. This is, of course, an understatement: the concept of vibration is the basis for numerous cosmologies of sound. See Marcus Boon, *The Politics of Vibration: Music as a Cosmopolitical Practice* (Duke University Press, 2022).

2. N. Katherine Hayles, *How We Became Posthuman: Virtual Bodies in Cybernetics, Literature, and Informatics* (University of Chicago Press, 1999), xiv.

3. Hayles, *How We Became Posthuman*; Kodwo Eshun, *More Brilliant than the Sun: Adventures in Sonic Fiction* (Quartet Books, 1998), 2.

4. Dempster, quoted in David W. Bernstein, *The San Francisco Tape Music Center: 1960s Counterculture and the Avant-Garde* (University of California Press, 2008), 263.

5. Curtis Roads and Morton Subotnick, "Interview with Morton Subotnick," *Computer Music Journal* 12, no. 1 (1988): 9–18.

6. Pauline Oliveros, *Software for People: Collected Writings 1963-80* (Smith Publications; Printed Editions, 1984).

7. Ra's Minimoog Model B did appear at a handful of his performances throughout the 1970s, such as his apparently disastrous performance at the Second World Festival of Black and African Arts and Culture (FESTAC) in 1977. But Ra largely favored instruments such as the Crumar Mainman, heard on dozens of recordings of live performances and studio albums from the mid-1970s to the 1990s.

8. Richard Alpert et al., "Rationale of the Mexican Psychedelic Training Center," in *Utopiates: The Use & Users of LSD 25*, ed. Richard Blum (Atherton Press, 1964), 179.

9. Ram Dass, *Be Here Now* (Hanuman Foundation, 1971), 5.

10. Dass, *Be Here Now*, 8.

11. Donald Buchla and Charles MacDermed, "The Electric Music Box," *Synthesis*, January 1971. RMP, box 32.

12. For more on Lilly, see Hannah Zeavin and Jeffrey Mathias, eds., *Reconsidering John C. Lilly* (MIT Press, forthcoming).

13. Pauline Oliveros, "Divisions Underground," in *Software for People: Collected Writings 1963–1980* (Smith Publications; Printed Editions, 1984), 97–111.

14. Pauline Oliveros, "Sonic Meditations," in *Source: Music of the Avant-Garde, 1966–1973*, ed. Larry Austin et al. (University of California Press, 2011), 342–46, http://www.jstor.org.remote.baruch.cuny.edu/stable/10.1525/j.ctt1png6w.73.

15. Pauline Oliveros, *Sonic Meditations* (Smith Publications, 1974).

16. See Nick Seaver, *Computing Taste: Algorithms and the Makers of Music Recommendation* (University of Chicago Press, 2022).

17. T. J. Pinch and Frank Trocco, *Analog Days: The Invention and Impact of the Moog Synthesizer* (Harvard University Press, 2002), 317–20.

18. Patriarchal descriptors began to be attached to both Buchla and Moog as early as the 1980s. See Dominic Milano, "American Synthesizer Builders: Triumphs and Crises for an Industry in Transition," *Keyboard*, May 1988. In recent years, "heritage brands" such as Buchla and Moog have both traded hands several times, which I have written about elsewhere. See Theodore Gordon, "The Buchla Music Easel: From Cyberculture to Market Culture," in *Modular Synthesis: Patching Machines and People*, ed. Ezra J. Teboul, Andreas Kitzmann, and Einar Engström (Routledge, 2024).

19. Fredric Jameson, "Periodizing the 60s," *Social Text*, nos. 9/10 (1984), 207.

20. Jameson, "Periodizing the 60s," 208.

21. The composer and improviser Sarah Belle Reid, for example, introduces the Buchla system housed at Mills College—the same system Buchla first built for the SFTMC—as such a tool. See "Exploring the 1st Buchla 100 Modular Synthesizer," posted February 2, 2024, https://www.youtube.com/watch?v=CiqhM-bNBrY&ab_channel=SarahBelleReid.

22. This is Seth Brodsky's translation in "'The Child at the Piano, Fumbling': A Hauskonzert," *Perspectives of New Music* 55, no. 1 (2017): 145. Original text is found in Theodor Adorno, "Improvisationen," in *Gesammelte Schriften*, Band 16, *Musikalische Schriften 1–3* (Suhrkamp, 1970), 263.

23. Theodor Adorno, Ästhetische Theorie, in *Gesammelte Schriften*, Band 7, *Ästhetische Theorie* (Suhrkamp, 2017), 55–56. This is Brodsky's translation in "Child at the Piano," modified from Theodor Adorno, *Aesthetic Theory*, ed. Gretel Adorno and Rolf Tiedemann, trans. Robert Hullot-Kentor (University of Minnesota Press, 1997), 32.

24. This is Brodsky's translation in "'Child at the Piano, Fumbling,'" modified from Adorno, *Aesthetic Theory*, 32.

25. This is the common English translation of Edouard Hanslick's 1865 phrase *tönend bewegte Formen*. For an in-depth discussion of the consequences of translating this phrase into English, see Nicole Grimes, "Eduard Hanslick's 'On the Musically Beautiful': A New Translation," *Musicologica Austriaca: Journal for Austrian Music Studies*, no. 2019 (September 2, 2019), https://www.musau.org//parts/neue-article-page/view/61.

26. Fumi Okiji, *Jazz As Critique: Adorno and Black Expression Revisited* (Stanford University Press, 2018), 5.

27. Okiji, *Jazz As Critique*, 6.

28. Fred Moten, *In the Break: The Aesthetics of the Black Radical Tradition* (University of Minnesota Press, 2003), 7.

Sources Cited

ARCHIVAL SOURCES

AACSR—Alton Abraham Collection of Sun Ra, University of Chicago Library, Chicago, Illinois
ALP—Alvin Lucier Papers, New York Public Library for the Performing Arts, Music Division, New York, New York
The Bancroft Library, University of California, Berkeley, Berkeley, California
CAA—Creative Audio Archive, Experimental Sound Studio, Chicago, Illinois
CCM—Center for Creative Music Archive, Mills College at Northeastern University, Oakland, California
CPEMCA—Columbia Princeton Electronic Music Center Archive, Columbia University Music Library (unprocessed), New York, New York
DTP—David Tudor Papers, Getty Research Foundation, Los Angeles, California
POP-NYPL—Pauline Oliveros Papers, New York Public Library for the Performing Arts, Music Division, New York, New York
POP-UCSD—Pauline Oliveros Papers, University of San Diego Library, San Diego, California
RAC—Rockefeller Archive Center, Sleepy Hollow, New York
RMP—Robert Moog papers, Cornell University Library, Ithaca, New York
San Francisco Conservatory of Music Library & Archives, San Francisco, California

Vasulka Archive, online
Yale Oral History of American Music, New Haven, Connecticut

INTERVIEWS WITH THE AUTHOR

Ramon Sender, San Francisco, California, July 1, 2015
William Maginnis and Ramon Sender, San Francisco, California, July 5, 2015
Morton Subotnick, Cortlandt Manor, New York, August 19, 2015
Roger Linn, Berkeley, California, February 4, 2016
Donald Buchla, Berkeley, California, February 6, 2016
Anthony Martin, Brooklyn, New York, September 4, 2017
Robert Gluck, email correspondence, March 11, 2024

DISCOGRAPHY

Brandeis University Chamber Chorus and Alvin Lucier. *Extended Voices*. Odyssey 32 16 0155, 1967.
Early Improvisations with Pauline Oliveros, Loren Rush, Terry Riley, and Others. Other Minds, 1957.
Kirk, Roland. *Rip, Rig, and Panic*. Limelight Records LM 82024, 1965.
Lucier, Alvin. *Vespers and Other Early Works*. New World Records 80604-2, 2002.
Oliveros, Pauline. *Four Electronic Pieces: 1959–1966*. Sub Rosa SR 185, 2017.
Oliveros, Pauline. *Reverberations: Tape & Electronic Music 1961–1970*. Important Records IMPREC352, 2012.
Reich, Steve, Richard Maxfield, and Pauline Oliveros. *New Sounds in Electronic Music (Come Out/Night Music/I of IV)*. Music of Our Time. Odyssey 32 16 0160, 1967.
Sender, Ramon. *Worldfood VII*. Locust Music L55, 2006.
Sonic Arts Union. *Electric Sound*. Mainstream Records MS/5010, 1972.
Subotnick, Morton. *Silver Apples of the Moon*. Nonesuch Records H-71174, 1967.
Sun Ra and His Astro-Solar Infinity Arkestra. *My Brother the Wind*. El Saturn Records 521, 1970.
Sun Ra and His Intergalactic Infinity Arkestra. *My Brother the Wind Volume II*. El Saturn Records 523/SRA 2000, 1971.
Sun-Ra and His Astro Infinity Arkestra. *My Brother the Wind*. Vol. 1. Cosmic Myth Records CMR002, 2017. https://sunramusic.bandcamp.com/album/my-brother-the-wind-vol-1-cd-lp-digital.
Sun-Ra and His Astro Infinity Arkestra. *My Brother the Wind*. Vol. 2. Cosmic Myth Music SATURN 523, 2014. https://sunramusic.bandcamp.com/album/my-brother-the-wind-vol-2.

Sun Ra and His Intergalactic Research Arkestra. *It's After the End of the World.* MPS Records CRM 748, 1970.
Sun Ra and His Solar-Myth Arkestra. *Nuits de la Fondation Maeght.* Vol. 1. Shandar SR 10 001, 1971.
Sun Ra and His Solar-Myth Arkestra. *Nuits de la Fondation Maeght.* Vol. 2. Shandar SR 10 003, 1971.
Sun Ra and the Intergalactic Research Arkestra. *Live in London.* Transparency 0317, 2005.
The Tornados. "Telstar." Decca 45-F 11494, 1962.
Various Artists. *New Music for Electronic and Recorded Media.* 1750 Arch Records S-1765, 1977.

FILMOGRAPHY/VIDEOGRAPHY

Anonymous Artists of America. Kartemquin Films, 1970.
The Computer and the Mind of Man. Program 1, *Logic by Machine.* KQED-TV/National Educational Television and Radio Center, 1962. http://archive.org/details/ComputerAndTheMindOfManP1LogicByMachine.
The Computer and the Mind of Man, Program 3, "The Universal Machine." KQED-TV/National Educational Television and Radio Center, 1962. http://archive.org/details/ComputerAndTheMindOfManP3TheUniversalMachine.
Intergalactic Music: Rare Sun Ra. "Sun Ra 1971 Oakland, CA Pre European Tour Rehearsal." WKCR, May 22, 1995. Posted April 6, 2022. https://www.youtube.com/watch?v=QLnp_1cLNlA.
McClaren, Norman. *Pen Point Percussion.* National Film Board of Canada, 1951. https://www.nfb.ca/film/pen_point_percussion/.
New Music: Sounds and Voices from the Avant-Garde New York 1971. Michael Blackwood Productions, 1972.
No Ideas but in Things: The Composer Alvin Lucier. Wergo, 2013.
Psych-Out. American International Pictures, 1968.
Reid, Sarah Belle. "Exploring the 1st Buchla 100 Modular Synthesizer." Posted February 2, 2024. https://www.youtube.com/watch?v=CiqhM-bNBrY&ab_channel=SarahBelleReid.
Space Is the Place. Directed by John Coney, 1974.
Subotnick: Portrait of an Electronic Music Pioneer. Waveshaper Media, 2017.
Sun Ra: A Joyful Noise. Directed by Robert Mugge, 1980.
"Young People's Concert: 'Bach Transmogrified.'" CBS, broadcast April 27, 1969. Posted April 26, 2018. https://youtu.be/J8sS5NkADBE?si=84SXPg2eA_4RbVBD.

BIBLIOGRAPHY

Adams, Frank, and Burgin Mathews. *Doc: The Story of a Birmingham Jazz Man*. University of Alabama Press, 2012.
Adorno, Theodor W. *Aesthetic Theory*. Edited by Gretel Adorno and Rolf Tiedemann. Translated by Robert Hullot-Kentor. University of Minnesota Press, 1997.
Adorno, Theodor W. *Gesammelte Schriften*. Band 7, *Ästhetische Theorie*. Suhrkamp, 2017.
Adorno, Theodor W. *Gesammelte Schriften*. Band 16, *Musikalische Schriften 1-3*. Suhrkamp, 2017.
Alpert, Richard, and Sydney Cohen. *LSD*. New American Library, 1966.
Alpert, Richard, Timothy Leary, and Ralph Metzner. "Rationale of the Mexican Psychedelic Training Center." In *Utopiates: The Use & Users of LSD 25*, edited by Richard Blum. Atherton Press, 1964.
Artaud, Antonin. *The Theater and Its Double*. Grove Weidenfeld, 1958.
Ashley, Robert. "The Wolfman, for Amplified Voice and Tape." In *Source: Music of the Avant-Garde, 1966-1973*, edited by Larry Austin, Douglas Kahn, and Nilendra Gurusinghe. University of California Press, 2011.
Atkins, Ronald. "Sun Ra." *Guardian*, November 10, 1970. AACSR, box 3, folder 4.
"Now There Are 606 Clears." *The Auditor: The Journal of Scientology* 30 (1967).
Babbitt, Milton. "An Introduction to the R. C. A. Synthesizer." *Journal of Music Theory* 8, no. 2 (1964): 251-65.
Babbitt, Milton. "Past and Present Concepts of the Nature and Limits of Music (1961)." In *The Collected Essays of Milton Babbitt*, edited by Stephen Peles, Stephen Dembski, Andrew Mead, and Joseph N. Straus. Princeton University Press, 2003. https://www.jstor.org/stable/j.ctt7rfx5.15.
Babbs, Ken. Sky Pilot Club home page. Accessed October 26, 2015. https://www.skypilotclub.com/index.1.html.
Barad, Karen. "Diffracting Diffraction: Cutting Together-Apart." *Parallax* 20, no. 3 (July 3, 2014): 168-87. https://doi.org/10.1080/13534645.2014.927623.
Baraka, Amiri. "Jazzmen: Diz & Sun Ra." *African American Review* 29, no. 2 (1995): 249-55. https://doi.org/10.2307/3042302.
Barnett, Myron. "Interview with Ramon Sender, Tony Martin, Morton Subotnick," University of Cincinnati College-Conservatory of Music, WGUC-FM, Cincinnati, July 13, 1964, Center for Contemporary Music Archive at Mills College, CCM CD1.
Barrett, G. Douglas. *Experimenting the Human: Art, Music, and the Contemporary Posthuman*. University of Chicago Press, 2023.
Batteau, Dwight W. "How to Make a Robot Speak English." *Analog Science Fiction and Fact*, August 1964.

Batteau, Dwight W., and Peter R. Markey. *Man/Dolphin Communication*. Final Report. US Naval Ordinance Test Station, China Lake, California, December 15, 1966—December 13, 1967.

Baum, L. Frank. *The Patchwork Girl of Oz*. Chicago: Reilly & Lee, 1913.

Behrman, David. "Wave Train." In *Source*, edited by Larry Austin, Douglas Kahn, and Nilendra Gurusinghe. University of California Press, 2011.

Bell, Eamonn. "Cybernetics, Listening, and Sound-Studio Phenomenotechnique in Abraham Moles's *Théorie de l'information et Perception Esthétique* (1958)." *Resonance* 2, no. 4 (December 1, 2021): 523–58. https://doi.org/10.1525/res.2021.2.4.523.

Bernstein, David W. *The San Francisco Tape Music Center: 1960s Counterculture and the Avant-Garde*. University of California Press, 2008.

Beta, Andy. "Sun Ra and His Arkestra: *The Magic City/My Brother the Wind* Vol. 1." Pitchfork, October 16, 2017. https://pitchfork.com/reviews/albums/sun-ra-and-his-arkestra-the-magic-city-my-brother-the-wind-vol-1/.

Blau, Herbert. *The Impossible Theater: A Manifesto*. Macmillan, 1964.

Bley, Paul, and David Lee. *Stopping Time: Paul Bley and the Transformation of Jazz*. Véhicule Press, 1999.

Bode, Harald. "A New Tool for the Exploration of Unknown Electronic Music Instrument Performances." *Journal of the Audio Engineering Society* 9, no. 4 (October 1961): 264–69.

Boon, Marcus. *The Politics of Vibration: Music as a Cosmopolitical Practice*. Duke University Press, 2022. https://doi.org/10.1215/9781478023012.

Bowker, Geof. "How to Be Universal: Some Cybernetic Strategies, 1943–70." *Social Studies of Science* 23, no. 1 (1993): 107–27.

Brodsky, Seth. "'The Child at the Piano, Fumbling': A Hauskonzert." *Perspectives of New Music* 55, no. 1 (2017): 145–206. https://doi.org/10.1353/pnm.2017.0002.

Brody, Martin. "The Enabling Instrument: Milton Babbitt and the RCA Synthesizer." *Contemporary Music Review* 39, no. 6 (November 1, 2020): 776–94. https://doi.org/10.1080/07494467.2020.1863011.

Brown, Jayna. *Black Utopias: Speculative Life and the Music of Other Worlds*. Duke University Press, 2021.

Buchla, Donald, and Charles MacDermed. "The Electric Music Box." *Synthesis*, January 1971. RMP, 8629, box 32. Division of Rare and Manuscript Collections, Cornell University Library.

Buchla, Donald F. Interview by Torsten Schmidt. Red Bull Music Academy, 2007. http://www.redbullmusicacademy.com/lectures/don-buchla-passing-the-acid-test.

Buchla, Donald F. *The Modular Electronic Music System*. Buchla Associates, 1966. http://www.vasulka.org/archive/Artists1/Buchla,DonaldF/ModElectrMusicSys.pdf.

Burnett, D. Graham. "Adult Swim: How John C. Lilly Got Groovy (and Took the Dolphin with Him), 1958–1968." In *Groovy Science: Knowledge, Innovation, and American Counterculture*, edited by David Kaiser and W. Patrick McCray. University of Chicago Press, 2016. https://doi.org/10.7208/chicago/9780226373072.001.0001.

Cage, John. *Silence: Lectures and Writings*. Wesleyan University Press, 1973.

Campbell, Robert, Christopher Trent, and Robert Pruter. "From Sonny Blount to Sun Ra: The Chicago Years." 2021. Revised May 11, 2025. http://campber.people.clemson.edu/sunra.html.

Carr, Danielle. "Political Anatomies of the Cyborg: Liberal Subjects and Neural Engineering." PhD thesis, Columbia University, 2023. https://doi.org/10.7916/bsfr-y255.

Carr, Danielle. "Scientology's Lonely Turf War." Pioneer Works, April 20, 2021. https://pioneerworks.org/broadcast/scientology-psychiatry.

Castaneda, Carlos. *The Teachings of Don Juan: A Jaqui Way of Knowledge*. 3rd California ed. University of California Press, 2020. Originally published 2008. https://doi.org/10.1525/9780520351400.

Chadabe, Joel. *Electric Sound: The Past and Promise of Electronic Music*. Prentice Hall, 1997.

Christmas, Shane. "Silver Apples Interview." Perfect Sound Forever, October 2000. https://www.furious.com/perfect/silverapples.html.

Chude-Sokei, Louis Onuorah. *The Sound of Culture: Diaspora and Black Technopoetics*. Wesleyan University Press, 2016.

Chusid, Irwin. "Liner Notes to *Atlantis*, by Sun Ra & His Astro Infinity Arkestra." Enterplanetary Koncepts, 2014. https://sunramusic.bandcamp.com/album/atlantis.

Chusid, Irwin. "Liner Notes to *My Brother the Wind* Vol. I." Enterplanetary Koncepts, 2017.

Ciani, Suzanne. "Don Buchla Memorial by Suzanne Ciani." *Huffington Post* (blog), October 6, 2016. https://www.huffingtonpost.com/suzanne-ciani/-don-buchla-memorial-by-s_b_12378852.html.

Clarke, Bruce. "John Lilly, The Mind of the Dolphin, and Communication Out of Bounds." *Communication +1* 3, no. 8 (2014). https://doi.org/10.7275/R5RB72JG.

Clynes, Manfred E., and Nathan S. Kline. "Cyborgs and Space." *Astronautics*, September 1960, 26–76.

Commanday, Robert. "Composer's Forecast: A New Kind of Artistic Man." *San Francisco Examiner*, April 12, 1965. Box 25, folder 14. Pauline Oliveros Papers, UCSD Special Collections.

Cox, Christoph. "The Alien Voice: Alvin Lucier's North American Time Capsule 1967." In *Mainframe Experimentalism: Early Computing and the Foundations of the Digital Arts*, edited by Hannah Higgins and Douglas Kahn. University of California Press, 2012.

Cox, Christoph. "Beyond Representation and Signification: Toward a Sonic Materialism." *Journal of Visual Culture* 10, no. 2 (August 1, 2011): 145–61. https://doi.org/10.1177/1470412911402880.

Cox, Christoph. *Sonic Flux: Sound, Art, and Metaphysics*. University of Chicago Press, 2018.

Cunningham, Carl. "Light and Sound in Berkeley." *San Francisco Chronicle*, January 29, 1966.

D'Alessandro, Jill, and Colleen Terry. *Summer of Love: Art, Fashion, and Rock and Roll*. University of California Press, 2017.

Daniel, Eric D., C. Denis Mee, and Mark H. Clark. *Magnetic Recording: The First 100 Years*. John Wiley & Sons, 1998.

"Darius Milhaud Part II: Paris & California." National Educational Television: The Creative Person. KQED, 1965. San Francisco Bay Area Television Archive. https://diva.sfsu.edu/collections/sfbatv/bundles/206370.

Dass, Ram. *Be Here Now*. Hanuman Foundation, 1971.

Dean, Roger T. *The Oxford Handbook of Computer Music*. Oxford University Press, USA, 2009.

Delio, Thomas. *Contiguous Lines: Issues and Ideas in the Music of the '60's and '70's*. University Press of America, 1984.

Dewan, Edmond M. "Occipital Alpha Rhythm Eye Position and Lens Accommodation." *Nature* 214, no. 5092 (June 1967): 975–77. https://doi.org/10.1038/214975a0.

Dewar, Andrew Raffo. "Reframing Sounds: Recontextualization as Compositional Process in the Work of Alvin Lucier." *Leonardo Music Journal* 22 (2012).

Drabinski, John E. *Glissant and the Middle Passage: Philosophy, Beginning, Abyss*. University of Minnesota Press, 2019. https://doi.org/10.5749/j.ctvh4zj15.

Drott, Eric. "Music and the Cybernetic Mundane." *Resonance* 2, no. 4 (December 1, 2021): 578–99. https://doi.org/10.1525/res.2021.2.4.578.

Dumas, Henry. "An Interview with Sun Ra." *Hiram Poetry Review*, no. 3 (Fall–Winter 1967): 31–32. AACSR, box 1, folder 4.

Dunbar-Hester, Christina. "Listening to Cybernetics: Music, Machines, and Nervous Systems, 1950-1980." *Science, Technology & Human Values* 35 (2010): 113–39.

The East Hampton Star. "Robert A. Dhaemers, Artist." April 3, 2014. http://easthamptonstar.com/Obituaries/2014403/Robert-Dhaemers-Artist.

Einhorn, Richard. "Dolphins Challenge the Designer." *Electronic Design*, December 6, 1967.

Epstein, John, Lindsay Davidson, Robert Burne, Reiner Burger, and David Sawyer. *The Black Box: An Experiment in Visual Theatre*. Latimer New Dimensions 2, 1970.

Erickson, Robert. *Sound Structure in Music.* University of California Press, 1975.
Eshun, Kodwo. *More Brilliant Than the Sun: Adventures in Sonic Fiction.* Quartet Books, 1998.
Fallon, Michael. "Bohemia's New Haven." *San Francisco Examiner*, September 7, 1965.
Fink, Bruce. *The Lacanian Subject: Between Language and Jouissance.* Princeton University Press, 1997.
Fiofori, Tam. "The Illusion of Sun Ra." *Liberator* 7, no. 12 (December 1967): 13–15. AACSR, box 1, folder 5.
Fiofori, Tam. "Moog Modulations: A Symposium." *DownBeat*, July 23, 1970.
Fiofori, Tam. "Myth, Music, and Media." *Glendora Review: African Quarterly on the Arts* 3, nos. 3 & 4 (2004): 101–14.
Fiofori, Tam. "Sun Ra & Africa." *BookArtVille* (blog), May 23, 2020. https://bookartville.com/sun-ra-africa/.
Fiofori, Tam. "Sun Ra's African Roots." *Melody Maker*, February 12, 1972.
Frankenstein, Alfred. "Subotnick's 'New' Music: 'An Heroic Vision' Is All of That." *San Francisco Chronicle*, September 26, 1962.
Fromm Music Foundation. "Princeton Seminars (1959 & 1960)." Accessed April 3, 2024. https://frommfoundation.fas.harvard.edu/princeton-seminars.
Galison, Peter. "The Ontology of the Enemy: Norbert Wiener and the Cybernetic Vision." *Critical Inquiry* 21, no. 1 (October 1994): 228–66. https://doi.org/10.1086/448747.
Gamer, Carlton. "Milton at the Princeton Seminars: A Remembrance." *Perspectives of New Music* 49, no. 2S (2012): 361–64. https://doi.org/10.7757/persnewmusi.49.2s.0361.
Ginsberg, Allen. "First Party at Ken Kesey's with Hell's Angels." In *Planet News: 1961–1967.* City Lights Publishers, 2001.
Glissant, Édouard. *Carribean Discourse: Selected Essays.* University of Virginia, 1999.
Glissant, Édouard. *Poetics of Relation.* Translated by Betsy Wing. University of Michigan Press, 1997.
Gluck, Bob. "Nurturing Young Composers: Morton Subotnick's Late-1960s Studio in New York City." *Computer Music Journal* 36, no. 1 (2012): 65–80.
Gluck, Bob. "Paul Bley and Live Synthesizer Performance." *Jazz Perspectives* 7, no. 3 (December 1, 2013): 303–22. https://doi.org/10.1080/17494060.2014.912257.
Goodrich, Juliette, and Molly McCrea. "Repair of Iconic '60s Era Synthesizer Turns into Long, Strange Trip for Engineer." CBS News Bay Area, May 21, 2019. https://www.cbsnews.com/sanfrancisco/news/1960s-rock-music-san-francisco-lsd-buchla-100-synthesizer-grateful-dead/.
Gordon, Theodore. "The Buchla Music Easel: From Cyberculture to Market Culture." In *Modular Synthesis: Patching Machines and People*, edited by Ezra J. Teboul, Andreas Kitzmann, and Einar Engström. Routledge, 2024.

Gottlieb, Lou. "The World of Music: Gloroius Electricity." *San Francisco Chronicle*, January 18, 1966.
Greenfield, Robert. "Owsley Stanley: The King of LSD." *Rolling Stone*, March 14, 2011. https://www.rollingstone.com/feature/owsley-stanley-the-king-of-lsd-82181/.
Greenfield, Robert. "Sun Ra & Europe's Space Music Scene." *Rolling Stone*, January 21, 1971.
Grimes, Nicole. "Eduard Hanslick's 'On the Musically Beautiful': A New Translation." *Musicologica Austriaca: Journal for Austrian Music Studies*, no. 2019 (September 2, 2019). https://www.musau.org//parts/neue-article-page/view/61.
Haraway, Donna. "The Promises of Monsters: A Regenerative Politics for Inappropriate/d Others." In *Cultural Studies*, edited by Lawrence Grossberg, Cary Nelson, and Paula A. Treichler. Routledge, 1992.
Hartman, Saidiya V. *Scenes of Subjection: Terror, Slavery, and Self-Making in Nineteenth-Century America*. Race and American Culture. Oxford University Press, 1997.
Haworth, Christopher. "Music and Cybernetics in Historical Perspective." *Resonance* 2, no. 4 (December 1, 2021): 461–74. https://doi.org/10.1525/res.2021.2.4.461.
Hayles, N. Katherine. *How We Became Posthuman: Virtual Bodies in Cybernetics, Literature, and Informatics*. University of Chicago Press, 1999.
Heinemann, Stephen. "Pitch-Class Set Multiplication in Theory and Practice." *Music Theory Spectrum* 20, no. 1 (1998): 72–96. https://doi.org/10.2307/746157.
Helmreich, Stefan. "Potential Energy and the Body Electric: Cardiac Waves, Brain Waves, and the Making of Quantities into Qualities." *Current Anthropology* 54, no. S7 (October 1, 2013): S139–48. https://doi.org/10.1086/670968.
Henahan, Donal. "Is Everybody Going to the Moog?" *New York Times*, August 24, 1969.
Hiller, Lejaren, and Leonard M. (Leonard Maxwell) Isaacson. *Experimental Music: Composition with an Electronic Computer*. McGraw-Hill, 1959.
Hinds, John. "Robert Moog Conversation with John Hinds—September 9, 1991—Stanford, Ca." Sun Ra Research, October 2000.
HKW Years of Now. "The Technological Big Bang: Tape Recorders, the Transistor & the Credit Card/Morton Subotnick." December 7, 2017. https://www.youtube.com/watch?v=Nvw4Z3Q0zaY&ab_channel=HKW100YearsofNow.
Hoffmann, E. T. A. "Beethoven's Instrumental Music: Translated from E. T. A. Hoffmann's 'Kreisleriana' with an Introductory Note." Translated by Arthur Ware Locke. *Musical Quarterly* 3, no. 1 (1917): 123–33.
Holmes, Thom. "Moog: A History in Recordings." June 17, 2013. https://moogfoundation.org/moog-a-history-in-recordings-by-thom-holmes-part-two/.
Holmes, Thom. "Sun Ra & the Minimoog." *The Bob Moog Foundation* (blog), November 6, 2013. https://moogfoundation.org/sun-ra-the-minimoog-by-historian-thom-holmes/.

Hopkins, Jerry, John Burks, and Paul Nelson. "The Groupies and Other Girls." *Rolling Stone*, February 15, 1969.

Huxley, Aldous. *Island*. Perennial Classic ed. New York: Perennial, 2009.

Iverson, Jennifer. *Electronic Inspirations: Technologies of the Cold War Musical Avant-Garde*. New Cultural History of Music. Oxford University Press, 2018.

Jacobs, Henry. "Vortex Comes Home—With Something New Added." *San Francisco Chronicle*, December 14, 1958.

Jameson, Fredric. "Periodizing the 60s." *Social Text*, nos. 9/10 (1984). https://doi.org/10.2307/466541.

Kahn, Douglas. *Earth Sound Earth Signal: Energies and Earth Magnitude in the Arts*. University of California Press, 2013.

Kahn, Douglas. "John Cage: Silence and Silencing." *Musical Quarterly* 81, no. 4 (1997): 556–98.

Kahn, Douglas. "The Military-Arts Nexus: Two Cases in the United States, c. 1970." *Studies in Material Thinking* 8 (May 2012). https://materialthinking.aut.ac.nz/sites/default/files/papers/SMT_V8_P02_Kahn_0.pdf.

Kampman, Lars. "Lars' Memoirs." Pittendrigh dis org. Accessed April 22, 2018. http://pittendrigh.org.

Keeling, Kara. *Queer Times, Black Futures*. New York University Press, 2019. https://www.jstor.org/stable/j.ctv12fw90q.

Kehew, Brian. "Liner Notes to *My Brother the Wind* Vol. I." Enterplanetary Koncepts, 2017.

Kettlewell, Ben. *Electronic Music Pioneers*. ProMusic Press, 2002.

Kirk, Roland. *Rip, Rig, and Panic*. Limelight, 1965.

Kline, Ronald R. *The Cybernetics Moment, or Why We Call Our Age the Information Age*. Johns Hopkins University Press, 2017.

Kreiss, Daniel. "Performing the Past to Claim the Future: Sun Ra and the Afro-Future Underground, 1954–1968." *African American Review* 45, nos. 1–2 (2012): 197–203. https://doi.org/10.1353/afa.2012.0006.

LaVigne, Robert. "Robert LaVigne: A Cyberspective." Big Bridge 6. Accessed March 14, 2018. http://www.bigbridge.org/issue6/lavigne.htm.

"Let It Sing!" *Time* 81, no. 23 (June 7, 1963): 83.

Lewis, George E. "Interacting with Latter-Day Musical Automata." *Contemporary Music Review* 18, no. 3 (January 1999): 99–112. https://doi.org/10.1080/07494469900640381.

Lewis, George E. "Too Many Notes: Computers, Complexity and Culture in Voyager." *Leonardo Music Journal* 10, no. 1 (December 1, 2000): 33–39.

Lilly, John C. *The Mind of the Dolphin*. Doubleday, 1967.

Listening, Inc. "Annual Report." *Analog Science Fiction and Fact*, July 1967.

Littlefield, Melissa M. *Instrumental Intimacy*. Johns Hopkins University Press, 2018. https://doi.org/10.1353/book.57783.

Lucier, Alvin. "Alvin Lucier—Thoughts on Installations." Kunstradio—Radiokunst Online. Accessed August 7, 2024. https://www.kunstradio.at/ZEITGLEICH/CATALOG/ENGLISH/lucier-e.html.
Lucier, Alvin. *Chambers: Scores by Alvin Lucier*. Wesleyan University Press, 1980.
Lucier, Alvin. *Reflections: Interviews, Scores, Writings/Reflexionen: Interviews, Notationen, Texte*. Edition MusikTexte 003. MusikTexte, 1995.
Maginnis, William. Interview by Eric Chasalow and Barbara Cassidy, June 27, 1997. The Video Archive of Electronic Music. http://ericchasalow.com/oralhist.html.
Masco, Joseph. "Nuclear Technoaesthetics: Sensory Politics from Trinity to the Virtual Bomb in Los Alamos." *American Ethnologist* 31, no. 3 (August 2004): 349–73. https://doi.org/10.1525/ae.2004.31.3.349.
Mathews, Burgin. "Sun Ra in Birmingham: A Few Ear(Th)Ly Artifacts." *Burgin Mathews* (blog), May 22, 2020. https://burginmathews.com/2020/05/22/sun-ra-in-birmingham-a-few-earthly-artifacts/.
McClure, Michael. "The Flowers of Politics, I." In *The New American Poetry: 1945–1960*, by Donald Allen. University of California Press, 1960.
McKeon, Richard P., David B. Owen, and Joanne K. Olson, eds. *On Knowing– The Social Sciences*. University of Chicago Press, 2017.
McKittrick, Katherine. "Axis, Bold as Love: On Sylvia Wynter, Jimi Hendrix, and the Promise of Science." In *Sylvia Wynter*, edited by Katherine McKittrick. Duke University Press, 2015. https://doi.org/10.1215/9780822375852-006.
McLuhan, Marshall. *Report on Project in Understanding New Media*. National Association of Educational Broadcasters, June 30, 1960. http://blogs.ubc.ca/nfriesen/files/2014/11/McLuhanRoPiUNM.pdf.
McLuhan, Marshall. *Understanding Media: The Extensions of Man*. Gingko Press, 2013.
The MEMS Project. "Touchplates." Accessed February 28, 2024. https://www.memsproject.info/touchplates.
Milano, Dominic. "American Synthesizer Builders: Triumphs and Crises for an Industry In Transition." *Keyboard*, May 1988.
Miles, Barry. *Hippie*. Sterling Publishing, 2005.
Miller, Brian A. "Leonard Meyer's Theory of Musical Style, from Pragmatism to Information Theory." *Resonance* 2, no. 4 (December 1, 2021): 475–502. https://doi.org/10.1525/res.2021.2.4.475.
Mills, Mara. "On Disability and Cybernetics: Helen Keller, Norbert Wiener, and the Hearing Glove." *Differences* 22, no. 2–3 (December 1, 2011): 74–111. https://doi.org/10.1215/10407391-1428852.
Minh-Ha, Trinh T. "She, the Inappropriate/d Other: Introduction." *Discourse* 8 (Fall–Winter 1986): 3–10.
Mockus, Martha. *Sounding Out: Pauline Oliveros and Lesbian Musicality*. Routledge, 2008.

Moeser, William. "Whiz Kid, Hands Down." *Life*, September 14, 1962.
Morton, David. *Sound Recording: The Life Story of a Technology*. JHU Press, 2006.
Moshier, Ross. Letter to Oliver Field, September 4, 1962. https://archive.org/details/fda.files/01.%20Folder1/page/n479/mode/2up.
Moshier, Stephen, William Van Lennep, and Edmund Perry. Acoustic Locator. 3558822. Cambridge, MA, filed May 22, 1968, and issued January 26, 1971.
Moten, Fred. *In The Break: The Aesthetics of the Black Radical Tradition*. University of Minnesota Press, 2003.
Mumma, Gordon, and Michelle Fillion. *Cybersonic Arts: Adventures in American New Music*. Music in American Life. University of Illinois Press, 2015.
Nakai, You. "On the Instrumental Natures of David Tudor's Music." PhD diss., New York University, 2016.
Nakai, You. *Reminded by the Instruments: David Tudor's Music*. Oxford University Press, 2021.
Nelson, Jerry Hopkins, John Burks, and Paul Nelson. "The Groupies and Other Girls." *Rolling Stone*, February 15, 1969.
Nye, David E. *America as Second Creation: Technology and Narratives of New Beginnings*. MIT, 2004.
O'Brien, Kerry. "Listening as Activism: The 'Sonic Meditations' of Pauline Oliveros." *New Yorker*, December 9, 2016. https://www.newyorker.com/culture/culture-desk/listening-as-activism-the-sonic-meditations-of-pauline-oliveros.
Okiji, Fumi. *Jazz As Critique: Adorno and Black Expression Revisited*. Stanford University Press, 2018. http://ebookcentral.proquest.com/lib/cunygc/detail.action?docID=5407254.
Oliveros, Pauline. "David Tudor: Testimonials," August 19, 1996. http://davidtudor.org/Life/testimonials.html.
Oliveros, Pauline. "Divisions Underground." In *Software for People: Collected Writings 1963–1980*. Smith Publications; Printed Editions, 1984.
Oliveros, Pauline. "Improvising Composition: How to Listen to the Time Between." In *Negotiated Moments: Improvisation, Sound, and Subjectivity*, edited by Gillian H. Siddall and Ellen Waterman. Improvisation, Community, and Social Practice. Duke University Press, 2016.
Oliveros, Pauline. Interview by Hannah Bächer. Red Bull Music Academy, 2016. https://www.redbullmusicacademy.com/lectures/pauline-oliveros-lecture.
Oliveros, Pauline. "Quantum Improvisation: The Cybernetic Presence." Keynote address presented at the conference Improvisation Across Borders at UCSD, April 11, 1999. https://www.hz-journal.org/n16/oliveros.html.
Oliveros, Pauline. "Reverberations: Eight Decades." *Jefferson Journal of Science and Culture*, no. 2 (2012). http://journals.sfu.ca/jjsc/index.php/journal/article/view/12.
Oliveros, Pauline. *Software for People: Collected Writings 1963–1980*. Smith Publications; Printed Editions, 1984.

Oliveros, Pauline. "Some Sound Observations." In *Software for People: Collected Writings 1963–1980*. Smith Publications; Printed Editions, 1984.
Oliveros, Pauline. *Sonic Meditations*. Smith Publications, 1974.
Oliveros, Pauline. "Sonic Meditations." In *Source: Music of the Avant-Garde, 1966–1973*, edited by Larry Austin, Douglas Kahn, and Nilendra Gurusinghe, 342–46. University of California Press, 2011. http://www.jstor.org.remote.baruch.cuny.edu/stable/10.1525/j.ctt1png6w.73.
Oliveros, Pauline. *Sounding the Margins: Collected Writings 1992–2009*. Deep Listening Publications, 2010.
Oliveros, Pauline. "Tape Delay Techniques for Electronic Music Composers." *Composer* 1, no. 3 (December 1969).
Oliveros, Pauline. "Tripping on Wires: The Wireless Body; Who Is Improvising?" *Critical Studies in Improvisation/Études Critiques En Improvisation* 1, no. 1 (2004). http://www.criticalimprov.com/article/view/9/31.
Oliveros, Pauline. "What Matters? Make the Music!" eContact!, September 2015. http://econtact.ca/17_3/oliveros_music.html.
Oliveros, Pauline, Stuart Dempster, John Luther Adams, and Monique Buzzarté. *Sounding the Margins: Collected Writings 1992–2009*. Edited by Lawton Hall. Deep Listening Publications, 2010.
Pareles, Jon. "Don Buchla, Inventor, Composer and Electronic Music Maverick, Dies at 79." *New York Times*, September 17, 2016. https://www.nytimes.com/2016/09/18/arts/music/don-buchla-dead.html.
Patterson, Orlando. *Slavery and Social Death: A Comparative Study*. Harvard University Press, 1985.
Patteson, Thomas. *Instruments for New Music: Sound, Technology, and Modernism*. University of California Press, 2015.
Peters, John Durham. "'Memorable Equinox': John Lilly, Dolphin Vocals, and the Tape Medium." *Boundary 2* 47, no. 4 (2020): 1–24. https://doi.org/10.1215/01903659-8677814.
Pickering, Andrew. *The Cybernetic Brain: Sketches from Another Future*. University of Chicago Press, 2010.
Pickering, Andrew. *The Mangle of Practice: Time, Agency, and Science*. University of Chicago Press, 1995.
Piekut, Benjamin. "Indeterminacy, Free Improvisation, and the Mixed Avant-Garde: Experimental Music in London, 1965–1975." *Journal of the American Musicological Society* 67, no. 3 (December 2014): 769–824. https://doi.org/10.1525/jams.2014.67.3.769.
Pinch, T. J., and Frank. Trocco. *Analog Days: The Invention and Impact of the Moog Synthesizer*. Harvard University Press, 2002.
Pinch, Trevor. "Why You Go to a Piano Store to Buy a Synthesizer." In *Path Dependence and the Social Construction of Technology*, edited by R. Garud and P. Karnoe. LEA Press, 2001.

"Princeton Seminars (1959 &1960)." Accessed August 28, 2024. https://fromm foundation.fas.harvard.edu/princeton-seminars.

Quarnstrom, Lee. *When I Was a Dynamiter! Or, How a Nice Catholic Boy Became a Merry Prankster, a Pornographer, and a Bridegroom Seven Times.* Edited by Iris Berry. Punk Hostage Press, 2014.

R. A. Moog Co. "Electronic Music Composition-Performance Equipment Short Form Catalog 1967." 1967. https://moogfoundation.org/wp-content/uploads/1967-R.A.-Moog-Catalog.pdf.

Ra, Sun. *The Immeasurable Equation: The Collected Poetry and Prose.* Edited by James L. Wolf and Hartmut Geerken. Waitawhile, 2005.

Rehding, Alexander. "Of Sirens Old and New." In *The Oxford Handbook of Mobile Music Studies*, vol. 2. edited by Sumanth Gopinath and Jason Stanyek. Oxford University Press, 2014. https://doi.org/10.1093/oxfordhb/9780199913657.013.003.

Richardson, Peter. *No Simple Highway: A Cultural History of the Grateful Dead.* St. Martin's Press, 2015.

Rietbrock, Bernhard. *Alvin Lucier's Reflexive Experimental Aesthetics.* Wolke Verlag, 2022.

Roads, Curtis, and Morton Subotnick. "Interview with Morton Subotnick." *Computer Music Journal* 12, no. 1 (1988): 9–18.

Rockwell, John. "The Musical Meditations of Pauline Oliveros." *New York Times*, May 25, 1980.

Rosenblueth, Arturo. Norbert Wiener, and Julian Bigelow. "Behavior, Purpose and Teleology." *Philosophy of Science* 10, no. 1 (1943): 18–24.

Rosenboom, David. "Saturation in Multi-Media." *Continuum*, September 1968.

Roszak, Theodore. *The Making of a Counter Culture: Reflections on the Technocratic Society and Its Youthful Opposition.* University of California Press, 1995.

Rubin, Nathan. *John Cage and the Twenty-Six Pianos of Mills College: Forces in American Music from 1940 to 1990, a History.* Sarah's Books, 1994.

San Francisco Chronicle. "Electronic Music Program." January 27, 1966.

Seaver, Nick. *Computing Taste: Algorithms and the Makers of Music Recommendation.* University of Chicago Press, 2022.

Secretan, Luc. "Cadmium-Sulfide Cell Micrometeorite Detector." National Aeronautics and Space Administration, 1962. NASA Space Science Data Coordinated Archive. https://nssdc.gsfc.nasa.gov/nmc/experimentDisplay.do?id=1962-070A-05.

Selvin, Joel. *San Francisco: The Musical History Tour; A Guide to Over 200 of the Bay Area's Most Memorable Music Sites.* Chronicle Books, 1996.

Sender, Ramon. "Ambulancia en el desierto: Ramón Sender y el San Francisco Tape Music Center." Interview by Rubén Coll, September 9, 2016. Radio del Museo Reina Sofía. http://radio.museoreinasofia.es/ambulancia-desierto.

Sender, Ramon. "The San Francisco Tape Music Center: A Report," in *The San Francisco Tape Music Center: 1960s Counterculture and the Avant-Garde*, edited by David Bernstein. University of California Press, 2008.
Shakespeare, William. *King Lear*. Classic Books Company, 2001.
Siders, Harvey. "Oliver's New Twist." *DownBeat*, July 23, 1970.
Steinskog, Erik. *Afrofuturism and Black Sound Studies: Culture, Technology, and Things to Come*. Palgrave Studies in Sound. Springer International/Palgrave Macmillan, 2018. https://doi.org/10.1007/978-3-319-66041-7.
Stern, Gerd. "From Beat Scene Poet to Psychedelic Multimedia Artist in San Francisco and Beyond, 1948–1978." Interview by Victoria Morris Byerly, 1996. Regional Oral History Office, The Bancroft Library, University of California, Berkeley. https://archive.org/details/beatscenepoetoogerdrich/.
Straebel, Volker, and Wilm Thoben. "Alvin Lucier's Music for Solo Performer: Experimental Music Beyond Sonification." In *Collected Work: Organised Sound*, vol. 19. Cambridge University Press, 2014.
Strange, Allen. *Programming and Meta-Programming in the Electro-Organism: An Operating Directive for the Music Easel*. Buchla & Associates, 1974.
Subotnick, Morton. "The Electric Heat of Creativity—Remembering Donald Buchla (1937–2016)." New Music Box, October 26, 2016. http://www.newmusicbox.org/articles/the-electric-heat-of-creativity-remembering-donald-buchla-1937-2016/.
Subotnick, Morton. "Morton Subotnick." Interview by Todd L. Burns. Red Bull Music Academy, 2011. https://www.redbullmusicacademy.com/lectures/morton-subotnick-2011.
Subotnick, Morton. "Morton Subotnick on Making Your Own Creative Tools." Interview by T. Cole Rachel, June 29, 2017. https://thecreativeindependent.com/people/morton-subotnick-on-making-your-own-creative-tools/.
Subotnick, Morton. "Morton Subotnick: The Mad Scientist in the Laboratory of the Ecstatic Moment." Interview by Frank J. Oteri, October 1, 2013. https://nmbx.newmusicusa.org/morton-subotnick-the-mad-scientist-in-the-laboratory-of-the-ecstatic-moment/.
Subotnick, Morton, and Suzanne Ciani. "Music Pioneers: Suzanne Ciani and Morton Subotnick in Conversation." Interview by Frosty, June 24, 2016. Red Bull Music Academy. http://daily.redbullmusicacademy.com/2016/06/encounters-suzanne-ciani-morton-subotnick.
Szwed, John. *Space Is the Place: The Lives and Times of Sun Ra*. Duke University Press, 1997.
Tanne, Janice Hopkins. "Humphry Osmond." *BMJ : British Medical Journal* 328, no. 7441 (March 20, 2004): 713.
Taubman, Howard. "The Arts: A Critic's View; 1958 Brussels Exposition Set a Mark That New York Is Far from Matching." *New York Times*, April 22, 1964.

Théberge, Paul. *Any Sound You Can Imagine: Making Music/Consuming Technology.* Music/Culture. Wesleyan University Press/University Press of New England, 1997.

Thomas, J. C. "Sun Ra's Space Probe." *DownBeat*, June 13, 1968.

Thompson, Hunter S. *Hell's Angels: A Strange and Terrible Saga.* Ballantine Books, 1996.

Turner, Fred. *The Democratic Surround: Multimedia and American Liberalism from World War II to the Psychedelic Sixties.* University of Chicago Press, 2015.

Turner, Fred. *From Counterculture to Cyberculture: Stewart Brand, the Whole Earth Network, and the Rise of Digital Utopianism.* University of Chicago Press, 2006.

Vale, V., and Andrea Juno. *Incredibly Strange Music*, Vol. 2. RE/Search Publications; Subco, 1993.

Verbos, Mark. "Made from the Best Stuff on Earth." *Buchla Tech* (blog), July 24, 2014. http://buchlatech.blogspot.com/2014/07/made-from-best-stuff-on-earth.html.

Von Gunden, Heidi. *The Music of Pauline Oliveros.* Scarecrow Press, 1983. http://archive.org/details/musicofpaulineoloooovong.

Von Hilgers, Philipp. "The History of the Black Box: The Clash of a Thing and Its Concept." *Cultural Politics* 7, no. 1 (March 1, 2011): 41–58. https://doi.org/10.2752/175174311X12861940861707.

Weheliye, Alexander G. "'Feenin': Posthuman Voices in Contemporary Black Popular Music." *Social Text* 20, no. 2 (2002): 21–47.

Weheliye, Alexander G. *Feenin: R&B Music and the Materiality of BlackFem Voices and Technology.* Duke University Press Books, 2023.

Weingarten, E. *The Music of ONCE: Perpetual Innovation.* Lulu.com, 2008.

Wiener, Norbert. *Cybernetics: Or Control and Communication in the Animal and the Machine.* 2nd ed. MIT Press, 1961. Originally published 1948.

Wiener, Norbert. *The Human Use of Human Beings: Cybernetics and Society.* Discus Books/Avon, 1967. Originally published 1950–1954.

Wilderson, Frank B., III. *Afropessimism.* Liveright Publishing, 2021.

Williams, Richard. "Night of the Gianta." *Melody Maker*, November 7, 1970. AACSR, box 3, folder 4.

Williams, Richard. "Sun Ra Coming!" *Melody Maker*, October 10, 1970. AACSR, box 3, folder 4.

Williams, Richard. "Sun Ra, Queen Elizabeth Hall." *The Times*, November 10, 1970. AACSR, box 3, folder 4.

Williams, William Carlos. "Paterson," in *The Collected Earlier Poems of William Carlos Williams* (New Directions Books, 1938), 231–235. https://www.fadedpage.com/showbook.php?pid=20200926.

Wittje, Roland. "The Electrical Imagination: Sound Analogies, Equivalent Circuits, and the Rise of Electroacoustics, 1863–1939." In *Music, Sound, and the Laboratory from 1750–1980.* University of Chicago Press, 2013.

Wolf, Daniel James. "No Ideas but in Things—The Composer Alvin Lucier—Text by Daniel James Wolf." 2013. http://alvin-lucier-film.com/text_wolf.html.
Wolfe, Tom. *The Electric Kool-Aid Acid Test*. Bantam Books, 1999.
Wynter, Sylvia. "Unsettling the Coloniality of Being/Power/Truth/Freedom: Towards the Human, After Man, Its Overrepresentation–An Argument." *CR: The New Centennial Review* 3, no. 3 (2003): 257–337. https://doi.org/10.1353/ncr.2004.0015.
Young, Ben. "Liner Notes to *My Brother the Wind* Vol. I." Enterplanetary Koncepts, 2017.Youngquist, Paul. *A Pure Solar World: Sun Ra and the Birth of Afrofuturism*. University of Texas Press, 2016.
Zeavin, Hannah, and Jeffrey Mathias, eds. *Reconsidering John C. Lilly*. MIT Press, forthcoming.

Index

Abraham, Alton, 173, 183
accordion, 91–92, 103–105, 104*fig*, 110–11
acoustic beats, 110–12
acoustic engineers, 11
acoustic siren, 47
Actor's Workshop of San Francisco, 33, 36
Adams, Frank "Doc," 164
Adorno, Theodor, 199–200
AFCRC. *See* Air Force Cambridge Research Center (AFCRC)
Afrofuturism, 161–62
Afropessimism, 185; Sun Ra's pessimism, 7, 184–185, 200
Ahmed (mynah bird), 103–4, 104*fig*, 105
Air Force Cambridge Research Center (AFCRC), 127, 131–32
Allen, Marshall, 22, 168
Allport, Gordon, 41–42
Alpert, Richard, 75–76, 82–84, 194–95, 196
Altamont Speedway Free Festival (1969), 86–87
alterity, Sun Ra and, 182–87
ambiguity, valorizing of, 85
"America Needs Indians," 79
American New Music, 27–28, 31–32
American Pavilion, Expo '58, 33
Ampex, PD-10 tape duplication system, 58, 106, 107*fig*, 113–14; Model 350, 226n43; three-track tape recorder, 51, 57–58; 213n1

analog electronic computers, 74–75
analog revival, 198
Analog Science Fiction and Fact, 143–44
anechoic chamber, 96
animals, testing on, 55
Ann Arbor, Michigan, 147
Anonymous Artists of America (AAA), 83–84, 88, 195
ANS Synthesizer, 50
Apple Box (Oliveros), 109
Arkestra, 19, 22, 155–56; as cybernetic system of individual players, 178; European Tour (1970), 157–59; *Nuits de la Fondation Maeght*, 174–80, 175*fig*; physical and metaphysical discipline, 179–83. *See also* Ra, Sun
Aspen Music Festival, 30
Atkins, Ronald, 158
Atlantis (Ra), 165–66, 175
Atlas Eclipticalis with *Winter Music, Electronic Version* (Cage), 105, 136
atmospheres, Subotnick's work with, 33–37
audiotape *See* Oliveros, Pauline; San Francisco Tape Music Center (SFTMC); tape recorders
Aurelius, Marcus, 29
automatic control, Buchla Box, 26

Babbitt, Milton, 11, 30–32, 45, 47
Babbs, Ken, 80

263

Bach, J.S., 165, 166
Barad, Karen, 129
Baraka, Amiri, 162–63
Barrett, G. Douglas, 16
Barron, Bebe, 12
Batteau, Dwight Wayne, 143–44
Baum, L. Frank, 94
Beaver, Paul, 167, 168
Be Here Now (Alpert/Dass), 194–95
Behrman, David, 12, 138
Belar, Herbert, 11
Bell, Eamonn, 11
Bell Labs, 38–40, 38*fig*
Beranek, Leo, 96
Berlin Jazz Festival, 177
Bernstein, David, 87
Bernstein, Leonard, 2, 167
Big Brother and the Holding Company, 78, 79–80, 79*fig*
Billboard, 1
biocomputer, 88–89, 140–43, 147, 150, 196
biofeedback, 12, 132–33
bioinformatics, 55
biological metaphors for music, 96–97
biometrics, black boxes and, 53, 54–57
black box, 51–52; current state of, 197–201; cybernetics (self-governing systems) and, 8–9; defined, 8, 235n14; ex-composition and, 193–97; experimenting with, 17–18; human mind as, 48; human musicians, role of, 190–201; metaphor of, 5–8; opacity and, 183–87; posthumanity and, 14–16; race and colors of metaphorical boxes, 160–61; *Waiting for Godot* production as, 36–37. *See also* composer's black box
black-boxed musical instruments, 13
black-boxed musicality, 18–22
black-boxed musical thinking, 13
Black popular music, 15
Blau, Herbert, 33, 34, 36–37, 44
Bley, Paul, 169, 238n44
Bode, Harald, 63
Bode Ring Modulator, 116
Boulez, Pierre, 60
Bowker, Geoffrey, 10, 13
brain waves, 131
Brand, Stewart, 79
Brandeis University, 127, 130–32
Brecht, George, 103
Brodsky, Seth, 199
Brody, Martin, 30–31, 32
Brown, Earle, 98
Brown, Jayna, 179–80

bubos, 36
Buchla, Donald, 3, 22, 48, 51–54, 106, 160, 166, 191, 192, 214n11; Buchla Box design, 26; countercultural interests of, 54–55; early systems of, biometrics and cybernetics, 54–57; early work at SFTMC, 57–59; ecstatic experience and, 195–97; Lucier, comparisons with, 154–55; ORB (Optical Ranging for the Blind), work on, 56, 59; philosophy of technology, 88–90, 195–96; psychedelics, use of, 77; Trips Festival, 79, 79*fig*; use of touch controlled voltage sources, 65–69, 66*fig*
Buchla Box, 3–4, 17–18, 19, 21, 61*fig*, 199; Anonymous Artists of America's box, 82–84; creation of, 27, 46–48; dual oscillators and sequential voltage sources, self-control and, 69–72, 70*fig*, 71*fig*; early versions of, 59–64, 61*fig*; ecstatic experience and, 195–97; ex-compositional uses of, 53, 81–87; as externalized model of human black box, 53–54; initial prototype, 52–53; interface of, 74–75, 207n61; Ken Kesey's Buchla Box, 81–82, 82*fig*; William Maginnis's *Flight*, 72–77, 73*fig*; Subotnick's vision of, 49–54; touch controlled voltage sources, 65–69, 66*fig*; transforming music from physical to informational endeavor, 3–4; Trips Festival, 77–81, 79*fig*; users' social identities and, 87–90
Buchla Box models: Model 107 Touch Controlled Voltage Source, 69; Model 111 Dual Ring Modulator, 73*fig*, 74–77; Model 112 Touch Controlled Voltage Source, 65–69, 66*fig*; Model 114 Touch Controlled Voltage Source, 68–69; Model 123 Sequential Voltage Source, 70–72, 71*fig*, 73*fig*, 74–77; Model 140 Timing Pulse Generator, 73*fig*, 74–77; Model 144 Dual Square Wave Generator, 69–72, 70*fig*; Model 146 Sequential Voltage Source, 71–72, 73*fig*, 74–77; Model 158 Dual Sine-Sawtooth Wave Generator, 69–72, 70*fig*; Model 160 White Noise Generator, 72–77, 73*fig*; Model 180 Dual Attach Generator, 73*fig*, 74–77
Buchla Music Easel, 17–18
Bye Bye Butterfly (Oliveros), 120–21

Cage, John, 16, 41, 43, 59, 96, 98, 103, 105, 132–33, 135; *Atlas Eclipticalis* with *Winter Music, Electronic Version*, 105, 136; *Cartridge Music*, 105; *0′00″*, 138
Calder, Alexander, 42

calibration, challenges with, 68
Callahan, Michael, 58
Carlos, Wendy, 1, 157–58, 166–67
Cartridge Music (Cage), 105
Chadabe, Joel, 72
Chamberlin Music Master, 60, 68
Chamber Music for Humans and Non-Humans (Lewis), 13
Chambers (Lucier), 129, 135
Chase, Ronald, 119
Chude-Sokei, Louis, 160–61, 187
Chusid, Irwin, 168
Ciani, Suzanne, 14, 88, 206n45
City Lights bookstore, 59
Clarke, Bruce, 145
Clavioline, 163, 164–65
Clynes, Manfred, 55
cognitive skills, computational intelligence and, 47–48
Cohen, Milton, 101
Columbia-Princeton Electronic Music Center (CPEMC), 11, 30–31
Columbia University, 11, 38–40, 38*fig*
Committee for National Morale, 41–42
"The Communication with Non-human Intelligences by Means of a Computer-Generated Musical Composition" (Lucier), 142–43
communist mind, 124–25
Company of Us (USCO), 58
composer, meaning of, 29–30
composer's black box, 5–8, 10, 191; Buchla's vision of, 52, 53; current state of, 197–201; human musicians, role of, 190–201; individual freedom and, 7–8; Lucier's vision of, 154; Oliveros's vision of, 108; Ra's vision of, 189; Subotnick, becoming a composer, 28–33, 46–48; Subotnick on black-boxed musicality, 18–23, 27–28, 50, 57, 60, 154–55, 189, 192–93; *See also* black box
Composition for Four Instruments (Babbitt), 30
Compton, Boyd, 78
computational theory of human mind, 141–42
computer music, 13
Computer Piece (Lucier), 150–53
The Computer and the Mind of Man (1962), 37–40, 38*fig*
Congo Club, Birmingham, 164
Conservatorio Santa Cecilia, 130
Consolidated Edison, 167
contact microphones, 21
contemporary posthuman, 16

counterculture, 54–55; Trips Festival, 77–81, 79*fig*. *See also* psychedelic expansion of the mind
Cowell, Henry, 47, 103
Cowell, Olive, 103
Cox, Christoph, 128, 134–35, 147
Coxe, Simeon III, 20, 207n62
CPEMC. *See* Columbia-Princeton Electronic Music Center (CPEMC)
Curtis, Eliot, 85–86
cybernetic body, 91–95; *Duo for Accordion and Bandoneon with Possible Mynah Bird Obbligato, See-Saw Version* (Oliveros), 102–8, 104*fig*, 107*fig*; improvising with tape, 97–101; Lucier and human biocomputer, 140–43; *Mnemonics* (Oliveros), 114–18, 117*fig*; *Mnemonics* tapes (Oliveros), 108–14, 109*fig*; *Music for Solo Performer* (Lucier), 12, 122, 128, 129, 130–35, 139, 151–52, 154; Oliveros and tape recorder as, 95–102; *SONICS* concert series and, 100–101; sound and cybernetic agency, 118–23; *Sound Patterns* (Oliveros), 102; *Time Perspectives* (Oliveros), 102; *Vespers* (Lucier) 129, 139, 140, 143–52, 196 *See also* biocomputer; on-line computational thinking
cybernetic organisms (cyborgs), 55
cybernetics, 51–52, 205n41; biofeedback, 132–33; biometrics, 54–57; Buchla Box design as, 26–28, 206n45; Buchla Box users, social identity and, 87–90; *The Computer and the Mind of Man* (1962), 38–40, 38*fig*; defined, 8–10; dual oscillators and sequential voltage sources, self-control and, 69–72, 70*fig*, 71*fig*; ecstatics and, 194–97; Erickson on musical composition as cybernetic system, 97; ex-composition and, 193–97; histories of enslavement in, 160–61; human mind as black box, 48; interdisciplinary use of language, 10; interpretation of *Music for Solo Performer* (Lucier), 134–35; Lucier and, 129; Maginnis's *Flight* with Buchla Box, 72–77, 73*fig*; music in America, 10–14; Oliveros's description of, 107–8; ontological theater and, 152; personality, definition of, 41–42; posthumanity and, 14–16; servomechanical enemy, cybernetic vision of, 125; Sun Ra performances and, 176–78; Sun Ra's cybernetic model of black box, 182–83
Cybernetics: or, Control and Communication in the Animal and the Machine (Wiener), 9
cybernetic sublime, 12–13, 20

dance of agency, 18
Dancers' Workshop, 42, 43
Darmstadt Summer Course, 30
Dass, Ram, 194–95
Davis, Danny, 168
De Blanc, Peter, 77–78, 214n11, 220n76, 220n77
"Deep Purple" (DeRose), 164
democratic surrounds, 41–42
Dempster, Stuart, 92, 192
DeRose, Peter, 164
Deutch, Herb, 63
Dewan, Edmond, 125–26, 127, 131, 132–33
Dewar, Andrew, 133–34
Dhaemers, Robert, 42, 44
dolphins, 140–46, 145*fig*
Donaueschingen Festival, 177
DownBeat, 1, 4, 169
Drabinski, John, 185–86
Drott, Eric, 11, 12, 20
Dumas, Henry, 185
Dunbar-Hester, Christina, 10
Duo for Accordion and Bandoneon (Oliveros), 103–6, 104*fig*
Durkee, Stephen, 58

Eaton, John, 63
echolocation, 143, 145, 146, 147–48, 149*fig*
ecstasis, state of, 83; black box, current state of, 198–201
ecstatics, science of, 76, 194–97
electroacoustic music, 13. *See also* specific artist names
electroencephalograms (EEGs), 21; electrical activity of brain, 131; Lucier's experiments with, 125–27, 126*fig*
electroencephalography, 134
electronic age, 50
electronic intermodulation, 111–12
electronic music, 13; abstract neutrality of, McLuhan and, 44–45; Subotnick's speculative instrument for, 50–51. *See also* specific artist names
Electronics Magazine, 167
Electronium (Scott), 20
elektronische Musik, 11, 51
enclosures, Subotnick's work with, 33–37
enemy behavior, 124–25
energy: flows between matter, 133–34; Moog synthesizers and, 2
♀ Ensemble, 119–20, 196–97
enslavement, experience of, 185–86
Environment and Sound Mobile, 42–45

Erickson, Robert, 93, 96–97, 100, 119; thinking sounds, 98–99
Eshun, Kodwo, 15–16, 17, 163
ex-composition, 75–76, 193–97; black box, current state of, 198–201; Buchla Box and, 81–87
experimental performance science, 145
Expo '58, American Pavilion, 33

Falkenstein, Claire, 97
Family Dog, 78
Farfisa combo organ, 159
fascism, 41–42
feedback, Buchla Box and, 26
feminism, Pauline Oliveros and, 119–21
filtering of electronic signals, 64
finger pressure, voltage control and, 67–68
Fiofori, Tam, 1, 2, 4, 161–62, 169, 172, 178–79
Flanagan, G. Patrick, 143–44
"The Flowers of Politics" (McClure), 34–35
Forbidden Planet (1956), 12
Foss, Lukas, 98
Frankenstein, Alfred, 34, 35–36
Frazer, Len, 84–85, 86*fig*, 88, 195
freedom: black box as free expression, 7–8; of individual performers, 98; Sun Ra's view of, 7, 161–62
free jazz, 179
Free Music Machine (Grainger), 50
Frith, Fred, 100
Fromm, Paul, 30
Furthur bus, 81–82, 82*fig*
Fylkingen Bulletin International, 135

Galison, Peter, 9, 55, 125
Gamer, Carlton, 32
Gate Hill Cooperative, Stony Point, New York, 40–41, 148–49, 149*fig*
Gaudeamus International Composers Award, 102
gender: ♀ Ensemble and, 119–21; posthumanity and, 15; posthuman vocality and, 15
General Telephone & Electronics Corporation, 124
gestural input, 25, 48, 49, 88; *Mnemonics* (Oliveros), 114–18, 117*fig*; *Music for Amplified Lip* (Lucier), 138–39, 152–153; Oliveros, experiments with HP 200CD oscillators, 112–14; Oliveros, patchwork improvisations, 101, 108, 118, 120; sound and cybernetic agency, 118–23; Sun Ra's use of Minimoog, 177, 188
Gibson/Kalamazoo 101 electric organ, 163, 165

Gilmore, John, 168
Glissant, Édouard, 185
global village, 41
Gluck, Bob, 75–76
Gnazzo, Anthony, 132
Golson, Benny, 169
Gottlieb, Lou, 80
Grainger, Percy, 50
Grateful Dead, 85
Great Grand Kludge, 57–59
Griffin, Donald, 147
Guardian, 158

Halprin, Anna, 42, 43
Happenings, 43–45
Haraway, Donna, 129
harmonic brotherhoods, Sun Ra and, 5–6
harmonics / harmonic distortion, 111, 113, 114
Harris, Elizabeth, 104–5, 119
Hartman, Saidiya, 188
Haworth, Christopher, 12
Hayles, N. Katherine, 14, 15, 16, 191–92
Heinemann, Stephen, 60
Hell's Angels, 86
Helmreich, Stefan, 56
Helms, Chet, 78–79
Hemsath, Bill, 171–72
Hiller, Lejaren, 11
Hinds, John, 170
Hip Hop, 15
Hoffmann, E.T.A., 96–97
Hohner Clavinet, 159
Hohner Electra, 159
HP 200CD oscillators, 108–14, 109*fig*
human body, 21. *See also* Oliveros, Pauline
human brain: biofeedback, 132–33; as biosignal instrument, 133–34; ecstatics and, 194–97; electrical activity of, 131; electrical potential of neurons as social potentials, 56; as musical instrument, 125–27, 126*fig*, 133–34
human consciousness, 52, 131
humanity: ex-composition and, 193–97; posthumanity, G. Douglas Barrett on, 16; posthumanity and black boxes, 14–17, 19, 191–92
human mind: as black box, 48, 125, 127; black box as model of, 13–14; Buchla Box users, social identity and, 87–90; cognitive skills, computational intelligence and, 47–48; computational theory of, 141–42; personality, definition of, 41–42; servomechanical enemy, cybernetic vision of, 124–25;

universal limitations of, 31. *See also* Oliveros, Pauline
The Human Use of Human Beings (Wiener), 160
Hunter, Meredith, 86–87
Huxley, Aldous, 103
hyperembodiment, 17

IBM, 38, 49, 81
Ichiyanagi, Toshi, 103, 105
ILLIAC computer, University of Illinois, 11
improvisation, 96–97; Oliveros, improvising with tape, 97–101; Sun Ra precise performances and, 159
information age, 6, 7–8
informational electronic signals, 131–32
information theory: application in music, 11–13; *The Computer and the Mind of Man* (1962), 38–40, 38*fig*; ecstatics and, 194–97; ideology of American New Music and, 31–32; Lucier and, 129; posthumanity and, 14–16
interface of Buchla Box, 74–75, 207n61
"Intergalactic Master" (Ra), 181
Intergalactic Research Arkestra. *See* Arkestra
intermodulation distortion, 111–12
International Federation for Internal Freedom, 83
intersubjective awareness, 180
Island (Huxley), 103
Iverson, Jennifer, 11

Jacopetti, Ben, 81
Jacopetti, Rain, 81
Jacson, James, 181–82, 184
Jameson, Frederic, 198–99
Jim Gurly, 79, 79*fig*
Johnson, Laurel, 99
Joplin, Janis, 78
Journal of the Audio Engineering Society, 63

Kahn, Douglas, 128, 131, 132–33
Kampman, Lars, 84, 85
Kaprow, Allan, 43, 44
Keeling, Kara, 188
Kesey, Ken, 80, 81–82, 82*fig*, 86–87, 88
Ketoff, Peter, 63
keyboards: Buchla Box, use of, 60; Buchla Box touch controlled voltage sources and, 65–69, 66*fig*; Clavioline, 163, 164–65; Gibson/Kalamazoo 101 electric organ, 163, 165; Minimoog, 163, 172–73, 175–77; Moog synthesizer, early uses, 168;

268 INDEX

keyboards (*continued*)
   Rock-Si-Chord, 163; Solar Sound Organ, 165–66; Solovox, 163, 164; Sun Ra's Minimoog Model B, 159–60
kinetic sculpture, 42
King Lear, 33–37
Kingsley, Gershon, 167–68, 169
Kirchner, Leon, 30, 33
Kirk, Rahsaan Roland, 20, 207n63
Kline, Nathan, 55
Kline, Ronald R., 6, 9, 38
Knight, Gloristeena (Ife Tayo), 158
KPFA radio station, 97, 99
KPIX television station, 100
KQED television station 37–40, 38*fig*, 43
Kreiss, Daniel, 178, 180
Krenek, Ernst, 30

Lacan, Jacques, 128
La Honda, 86
Lawrence Radiation Laboratory (Rad Lab), 51, 54–57
League of Automatic Music Composers, 13
Leary, Timothy, 83
Le Marteau sans Maître (Boulez), 60
lesbian musicality, 119–20
Leuning, Otto, 11
Lewis, George, 13
Lilly, John C., 140–43, 146, 196
Lincoln Center, 167
Lincoln Center Repertory Theater, 78
liquid light shows, 77–78
Listening, Inc., 143–46, 145*fig*
Listening in the Dark (Griffin), 147
Littlefield, Melissa, 134
Live Wire Entertainment, 164
Lloyd, Norman, 78
Longshoremen's Hall, 78–79
Lonidier, Lynn, 119
LSD (lysergic acid diethylamide), 52, 194–95; Buchla Box as electronic equivalent of, 76–77; Trips Festival, 77–81, 79*fig*
Lucier, Alvin, 12, 21, 22, 93, 122–130, 126fig, 149fig, 160, 191–93, 196; black-boxed musicality, 130; *Chambers*, 129, 135; "The Communication with Non-human Intelligences by Means of a Computer-Generated Musical Composition," 142–43; comparisons with Buchla, Oliveros, and Subotnick, 154–55; *Computer Piece*, 150–51, 152–53; echolocation, visual analogs to, 147–48; human biocomputer and, 140–43; interpretations of early works, 127–30; *Music for Amplified Lip*, 136–39, 152–53; *Music for Solo Performer*, 12, 122, 128, 129, 130–35, 139, 151–52, 154; *Music on a Long Thin Wire*, 151; "no ideas but in things" as motto, 151–56; ONCE Festival, 147; postwar technoscience, entanglement with, 130–32; *Signatures*, 136–37, 152–53; sketches (1965–66), 135–39; *Vespers*, 129, 139, 140, 143–52, 196
Lucier, Mary, 148–49, 149*fig*

MacDermed, Charles, 79, 79*fig*
Madama Butterfly (Puccini), 120–21
Magic Theater for Madmen Only, 77–81, 79*fig*
The Magic City (Ra), 165
Maginnis, William, 51, 53, 58, 72, 88, 106, 107*fig*; Buchla Box, first *Flight* with, 72–77, 73*fig*; psychedelics, use of, 77
magnetrons, 8
"Man to Dolphin Translator," 143–44
Marfa, Texas, 147
Martin, Anthony, 44, 60, 78, 79
Masco, Joseph, 37
Massachusetts Institute of Technology (MIT), 8–9, 11
McClure, Michael, 34
McKeon, Richard, 84–85
McLaren, Norman, 47
McLuhan, Marshall, 25, 28, 40–42, 43, 44–45, 50, 192
media theory, 28, 50
meditative state, 131–32
Meek, Joe, 164
Melody Maker, 157
Menlo Park, 84
Merry Pranksters, 80, 81–82, 82*fig*, 84, 86
metaphysical discipline, 180–83
meta-programming, 142
Metzner, Ralph, 83
Meyer, Leonard, 11
Meyer-Eppler, Werner, 10–11
Milhaud, Darius, 30, 42, 43
military-industrial-academic complex, 127, 129, 143–44, 194
Mills College, 42
Mills College Tape Music Center, 118–19, 132
The Mind of the Dolphin: A Nonhuman Intelligence (Lilly), 140–43, 146
Minh-Ha, Trinh T., 129
Minimoog, 5, 19–20, 162, 170–72, 171*fig*; *Nuits de la Fondation Maeght* (Ra),

174–80, 175*fig*; prototypes of, 171–72, 171*fig*; sounds produced by, 163; Sun Ra's Minimoog Model B, 159–60; Sun Ra's acquisition of, 169–74; Sun Ra, early performances with, 172–73; Sun Ra and, 155–56; Sun Ra's use of to demonstrate infinity, 181–82; Sun Ra's Minimoog Model D, 193
mise-en-scéne, 34
Mnemonics (Oliveros), 114–18, 117*fig*
Mockus, Martha, 119–21
Modular Electronic Music System, 26, 61*fig*, 182, 214n13, 218n54. *See also* Buchla Box
Moles, Abraham, 10–11
Moog, Robert, 1, 2, 53, 63, 64, 65, 162, 166–67, 169–72; composer's black box, vision of, 6–7; financial concerns of, 171–72
Moog synthesizer, 17, 19, 162; as black box, 5–8; conceptualization and design, 166–67; emergence of, 1–3; music as an informational endeavor, 3–4; prohibitive cost of, 168–69; as sculpting sound, 2–3; Sun Ra's acquisition of, 169–74; Sun Ra's use of, 157–60; Wendy Carlos and, 166–67. *See also Minimoog*
Moten, Fred, 186, 201
Mumma, Gordon, 138
Murzin, Yevgeny, 50
music: abstract neutrality of, McLuhan and, 44–45; black box, current state of, 197–201; black-boxed musicality, 18–22, 47–48; in cybernetic America, 10–14; deskilling and democratizing of musical creativity, 6; as studio art, 3–4, 18–19, 27; Sun Ra's concept of harmonic brotherhoods, 5–6; transforming from physical to informational endeavor, 3–4, 47–48
music easel, 5–6, 17–18, 27, 49–54. *See also* Buchla Box; Subotnick, Morton
Music for Amplified Lip (Lucier), 136–39, 152–53
Music for Solo Performer (Lucier), 12, 122, 128, 129, 130–35, 139, 151–52, 154
musicians: black-boxed musicality, 18–22; embrace of cybernetic ideas, 10–14; encounters with technoscience, 7–8. *See also* specific musician names
musicking human subjects, 13
"Music on a Long Thin Wire" (Lucier), 151
musique concrète, 51, 56–57
My Brother the Wind Volume II (Ra), 173
"My Music Is Words" (Ra), 183–84

National Aeronautics and Space Administration (NASA), 55
National Association of Educational Broadcasters, 40–42
National Defense Research Committee, 96
National Educational Television and Radio Center, 37–40, 38*fig*
Nature, 131
Nelson, Oliver, 169
Neurophone, 144
New York Times, 167
New York University, Intermedia program, 78
Nonesuch Records, 78
nonlinear systems, 9–10. *See also* cybernetics
nuclear sublime, 37
Nuits de la Fondation Maeght (Ra), 174–80, 175*fig*
NuMus West, 94, 196
Nye, David, 6

objet sonore, 11
O'Brien, Kerry, 119
Office of Naval Research, 127
Okiji, Fumi, 200–201
Olatunji, Babatunde, 165
Olatunji Center of African Culture, 165
Oliveros, Pauline, 17, 21, 22, 51, 72, 90, 160, 191–93; *Apple Box*, 109; body electric, 101–2; *Bye Bye Butterfly*, 120–21; cybernetic body, 91–95; *Duo for Accordion and Bandoneon with Possible Mynah Bird Obbligato, See-Saw Version*, 102–8, 104*fig*, 107*fig*; ecstatics and, 196–97; ♀ Ensemble, 119–20, 196–97; experimentation with oscillators, 91–93; first encounter with tape recorder, 95–102; improvisation, use of, 96–97, 103–4; improvising with tape, 97–101; Lucier, comparisons with, 154–55; *Mnemonics*, 114–18, 117*fig*; *Mnemonics* tapes, 108–14, 109*fig*; as Patchwork Girl, 94–95; patchwork metaphor, use of, 93–95; *Pieces of Eight*, 93, 106; psychedelics, rejection of, 77; "research of a patchwork quilt nature," 106–8, 107*fig*; sonic energy, explorations of, 122–23; *Sonic Meditations*, 119–20, 196–97; SONICS concert series, 100; sound and cybernetic agency, 118–23; *Sound Patterns*, 102; Subotnick, comparison with, 108; "Tape-A-Thon" concert, 119; thinking sounds, 98–99; *Time Perspectives*, 102; Tudorfest, 103–6, 104*fig*

INDEX

Olson, Harry, 11, 31
ONCE Festival, 147
on-line computational thinking, 149–51, 149*fig*, 195–96
On the Sensations of Tone (Helmholtz), 61
ontological theater, 133, 152–53
ontology of the enemy, 125
opacity, black box and, 183–87, 200–201
Open Theater, 81
operationalism, 84–85
Oppenheimer, Robert, 30
Oram, Daphne, 50
ORB (Optical Ranging for the Blind), 56, 59
organicist metaphors for music, 96–97
organic modernism, 42
oscillators, use of, 107*fig*; dual oscillators and sequential voltage sources, self-control and, 69–72, 70*fig*, 71*fig*; *Mnemonics* (Oliveros), 114–18, 117*fig*; Oliveros, HP 200CD oscillators and, 108–14, 109*fig*; Oliveros's experimentation with, 91–93; touch controlled voltage sources, 66–69, 66*fig*. *See also* Oliveros, Pauline
oscilloscopes, use of, 59, 62, 75
Oteri, Frank, 44
Oxford Handbook of Computer Music, 133–34

parascientific audio devices, 21
Patchwork Girl, 94
patchwork metaphor, Oliveros's work as, 93–95. *See also* Oliveros, Pauline
Patterson, Orlando, 185
"Patterson" (Williams), 151, 153–54
Paul, Les, 20
Payne, Maggi, 87
Perry Lane, 84
personality, 41–42
Peters, John Durham, 141
physical and metaphysical discipline, 179–83
Pickering, Andrew, 18, 133, 152
Pieces of Eight (Oliveros), 93, 106
Pinch, Trevor, 198
Polyester Moon (Falkenstein), 97
posthumanity, 191–92; black-boxed musicality and, 19; black boxes and, 14–16; contemporary posthuman, 16
postwar experimental music, 16
potentiometers, use of, 75
Princeton Seminar in Advanced Musical Study, 30, 31–32
Programming and Metaprogramming in the Electro-Organism (Strange), 17
propaganda, 41–42

psychedelic expansion of the mind, 21, 26–27, 194–95; Donald Buchla and, 52; excompositional uses of Buchla Box, 75–76, 81–87; Trips Festival, 77–81, 79*fig*
psychoacoustics, 111–12
Psych-Out (1968), 85
Puccini, Giacomo, 120

Quarnstrom, Lee, 84
Queen Elizabeth Hall, 157–59, 177–78

Ra, Sun, 4–5, 19–20, 22, 93, 98, 108, 122, 130, 155–56, 191–93; *Atlantis*, 165–66, 175; composer's black box, vision of, 7; "The Cosmic Explorer," 176–77; cybernetic model of black box, 182–83, 187–89; discipline and precision of, 158–59, 161–62, 179–80, 189, 241n110; early relationship to technology, 162–66; ecstatic state and, 197; European tour (1970), 157–59; excomposition and, 197; on freedom, 7, 200–201; "Intergalactic Master," 181; *The Magic City*, 165; Minimoog, early performances with, 172–73; Moog synthesizer, introduction to, 167–74; *My Brother the Wind*, 173, 237n41; *My Brother the Wind Volume II*, 173, 237n41, 238n47, 238n58; "My Music Is Words," 183–84; *Nuits de la Fondation Maeght*, 174–80, 175*fig*; ontology of "alter-life," 182–87; physical and metaphysical discipline, 179–83; refusal of earthly ways of knowing, 161–62; Saturn Records label, 164; scenes of subjection and, 188–89; *Space Is the Place*, 163, 193; "The Universe Sent Me to Converse with You," 187
race / racism: effect on views of freedom, 7; metaphorical boxes, colors of, 160–61; posthumanity and, 15; posthuman vocality and, 15; racialized Black box, 182–89; Robert Moog attitude toward Sun Ra and Paul Bley, 169–72; Sun Ra on the "alter-life," 182–89
Rancho Diablo, 84
RAND Corporation, 38–40, 38*fig*
RCA Mk. II Sound Synthesizer, 11, 30–31, 32, 45, 47–48
RCA Sarnoff Research Center, 30–31
real-time self-modification, Buchla Box and, 26
Report on Project in Understanding Media, McLuhan (1960), 40–42
Rhythmicon, 47

INDEX   271

Richards, M. C., 40–41, 148–49
Riley, Terry, 97
ring modulators, 111–12, 147
RMI Rock-Si-Chord, 159, 163
Rockefeller Foundation, 11, 78, 214n11, 216n33, 216–217n43, 220n76
Roland 808 Rhythm Composer, 15
Romero, Elias, 77
Rose Art Museum, 132
Rosenboom, David, 12
Roszak, Theodore, 54
Rubin, Nate, 30
Rush, Loren, 97
rusty bolt effect, 111–12, 113
Rzewski, Frederic, 130

Saint-Paul de Vence, France, 174
Sands Jr., Major General Harry James, 124–25
San Francisco Chronicle, 22–23, 49, 80–81
San Francisco Conservatory of Music (SFCM), 58, 96, 100, 102
San Francisco counterculture, 21
San Francisco Examiner, 57, 100
San Francisco Museum of Art, 40
San Francisco State College, 96
San Francisco Tape Music Center (SFTMC), 3, 21, 28, 40–42, 44, 46, 51, 102, 214n11; Buchla's arrival at, 57–59; Buchla's early work at, 57–59; liquid light shows, 77–78; move to Mills College, 118–19; *Mnemonics* (Oliveros), 108–14, 109*fig*; studio improvements, Oliveros's research and, 106–8, 107*fig*; Trips Festival, 77–81, 79*fig*
Saturday Review, 167
Saturn Records, 164
scenes of subjection, 188–89
Schaeffer, Pierre, 11
scientific data sonification, 133–34
Scientology, 143–44, 167, 231n69, 232n77
Scott, Jim, 171
Scott, Raymond, 20
Seashore, Carl, 31
self-governing systems, 8–9, 69–72, 70*fig*, 71*fig*. *See also* cybernetics
semiotic information, 150–51
Sender, Ramon, 17, 28, 40–42, 44, 46, 51, 58, 59–60, 68, 77, 79, 81, 100; Trips Festival, 79–80, 79*fig*
Sessions, Roger, 30
SFTMC. *See* San Francisco Tape Music Center (SFTMC)
Shannon, Claude, 11, 32

Siders, Harvey, 169
Signatures (Lucier), 136–37, 152–53
silent sounds, 96
Silver Apples (band), 20
Silver Apples of the Moon (Subotnick), 17, 19, 24–28, 53, 78
Simon, Douglas, 140
sine waves, acoustic beat frequencies and, 111–12; *Mnemonics* (Oliveros), 114–18, 117*fig*
Smart-Grosvenor, Verta, 158
Smith, Stuff, 164
Smythe, Henry D., 30
social death, 185
social identity, users of Buchla Box and, 87–90
Solar Sound Organ, 165–66. *See also* Gibson/Kalamazoo 101 electric organ
Solovox, 163, 164
Sondol, 127, 140, 143, 144–46, 145*fig*, 147, 148–50, 152, 196, 232n77, 233n78
Sonic Arts Group/Sonic Arts Union, 12
sonic flux, 128
sonic materiality, 151–52
Sonic Meditations (Oliveros), 119–20, 196–97
sonic naturalism, 147
sonic output, 48
SONICS concert series, 58, 100
sonic sculpture, 42
Sonny Blount's Solo Vox Band, 164
sound, as informational signal, 112
Sound Blocks: An Heroic Vision, 33–37, 44
Sound Patterns (Oliveros), 102
Sound Structure in Music (Erickson), 97
sources for book, 22–23
space chord, 158
Space Daisy, 84
Space Is the Place (Ra), 163, 193
Spacemaster organ, 159
spatialized speaker systems, 101
spectrographic analysis, 115–18, 117*fig*
Stanley, Augustus Owsley "Bear" III, 52, 77, 83, 85–86
Stern, Gerd, 40–41, 58
Stockhausen, Karlheinz, 11
Stoicism, 29–30
Stony Brook University, 116
Sträbel, Volker, 128, 139
studio art, music as, 3–4, 18–19, 27
Studio di Fonologica, 130
Subotnick, Morton, 3, 17, 19–20, 21, 22–23, 98, 160, 191–93, 195, 212–13n1, 214n11; becoming a composer, 28–33; Buchla Box,

Subotnick, Morton (*continued*)
user identity and, 88; characterization of synthesizer, 25–26; composer's black box, vision of, 5–8, 46–48, 49–54; *The Computer and the Mind of Man* (1962), 37–40, 38*fig*; departure from SFTMC, 78; *Environment and Sound Mobile*, 42–45; interpretation of Marshall McLuhan's media theory, 40–42; *King Lear* and *Sound Blocks*, 33–37, 44; Lucier, comparisons with, 154–55; music as studio art, 18–19; music easel, vision of, 59–64; Oliveros, comparison with, 108; psychedelics, rejection of, 77; *Silver Apples of the Moon*, 24–28, 53, 78

Sun Ra Research, 170
Swansen, Chris, 2–3, 4, 6–7
Switched-On Bach (Carlos), 1, 157–58, 166–67
Sylvania Electronics Systems, 124–126, 127; Sylvania Scanner, 124–30, 126*fig*, 132, 134; vocoder, 127–28
SynKet instrument, 63
synthesizers: current state of, 197–201; early models of, 63–64; emergence of, 1–2; Subotnick's characterization of, 25–26. *See also* specific instrument names
System, 167
systems thinking, operationalism and, 84–85
Szwed, John, 165, 174, 179–80

Tanaka, Atau, 133
"Tape-A-Thon" concert, 119, 120
tape recorders, 20, 33; Ampex, three-track tape recorder, 51, 57–58; Ampex PD-10 tape duplication system, 58, 106, 107*fig*, 113–14; Ampex 350, 226n43; challenges of using, 50–51; *Environment and Sound Mobile*, 42–45; Ken Kesey's Buchla Box, 82; *King Lear* and *Sound Blocks*, 33–37; Maginnis's *Flight* with Buchla Box, 72–77, 73*fig*; musique concrète, 56–57; sequencers, 58; Subotnick, creation of composer's black box, 46–48; Subotnick's enclosures and atmospheres, 34, 39; Viking tape cartridge machines, 68, 69; *See also* Oliveros, Pauline; San Francisco Tape Music Center (SFTMC); Sender, Ramon
Tartini tones, 111
technocracy, 54
technology/technoscience, 7–8; Buchla's philosophy of, 195–96; grammar of dominance, 160–61; ideology of American New Music and, 31–32; new age, conceptual power of, 37; nuclear sublime and, 37; Sun Ra's early relationship with technology, 162–66; utopian narratives about, 6, 20

technopoetics, 154
technoscience, 8, 129, 152, 192; Lucier and, 129–30, 135, 139, 151, 152, 182, 192; Oliveros and, 98, 182, 192; Subotnick and, 31–32, 37; Sun Ra and, 182, 192
"The Cosmic Explorer" (Ra), 176–77
The Electric Kool-Aid Acid Test (Wolfe), 80
Theremin, Leon, 47; theremin (instrument), 165
thinking sounds, 98–99
Thoben, Wilm, 128, 139
Thompson, Hunter S., 86
Time, 43
Time Perspectives (Oliveros), 102
Times, 158
Tinguely, Jean, 42
Toccatta and Fugue in D Minor (Bach), 165–66
totalitarianism, technology and, 54
touch controlled voltage sources, use of, 65–69, 66*fig*; decay time, 68–69; dual oscillators and sequential voltage sources, self-control and, 69–72, 70*fig*, 71*fig*
Trocco, Frank, 198
Tudor, David, 12, 93, 103, 104*fig*, 120, 130, 148–49
Tudorfest, 103–6, 104*fig*
Tukey, John, 32
Turner, Fred, 41–42, 54–55
Tyson, June, 158

ultrasonic signals, 147–148, 56, 110, 112
Understanding Media (McLuhan), 25, 40–42, 45
unified field of experience, McLuhan, 50
"The Universe Sent Me to Converse with You" (Ra), 187
University of Cincinnati College-Conservatory of Music, 28
University of Illinois, 11
University of San Diego (UCSD), 119–20
University of Toronto Electronic Music Studio (UTEMS), 118–19
US Army Signal Corps, 143
US Naval Ordinance Test Station (USNOTS), 143–44
Ussachevsky, Vladimir, 11

utopian technological narrative, 6, 20, 155–56, 198–201

Vargas, Fred, 167–68
very low frequency (VLF) radio antennas, 127
Vespers (Lucier), 129, 139, 140, 143–52, 196
Viking tape cartridge machines, 68, 69
Vivian Beaumont Theater, 78
vocoders, 15, 126
voltage sources, touch controlled, 65–69, 66*fig*; decay time, 68–69; dual oscillators and sequential voltage sources, self-control and, 69–72, 70*fig*, 71*fig*; Maginnis's *Flight* with Buchla Box, 72–77, 73*fig*
von Gunden, Heidi, 120
von Helmholtz, Hermann, 61, 64
von Hilgers, Philipp, 8
von Neumann, John, 32

Waiting for Godot, 36–37

Weheliye, Alexander, 15, 16
Weiner, Norbert, 55–56
Weiss, Jon, 170
West-Deutscher Rundfunk (WDR) studio, 11
white box, 160–61
Wiener, Norbert, 8–10, 11–12, 13, 14, 55, 131, 132–33, 160, 195, 235n14
Wilderson, Frank B. III, 185
Williams, Richard, 157–58
Williams, William Carlos, 151, 153–54
Wolf, Daniel L., 127
Wolfe, Tom, 80
Wolff, Daniel, 151–52
World Food series (Sender), 68
Wynter, Sylvia, 184–85

Xenakis, Iannis, 11
X-rays, hazard of, 59

Youngquist, Paul, 189

Founded in 1893,
UNIVERSITY OF CALIFORNIA PRESS
publishes bold, progressive books and journals
on topics in the arts, humanities, social sciences,
and natural sciences—with a focus on social
justice issues—that inspire thought and action
among readers worldwide.

The UC PRESS FOUNDATION
raises funds to uphold the press's vital role
as an independent, nonprofit publisher, and
receives philanthropic support from a wide
range of individuals and institutions—and from
committed readers like you. To learn more, visit
ucpress.edu/supportus.